beyond genetics

PUTTING THE POWER OF DNA
TO WORK IN YOUR LIFE

beyond genetics

PUTTING THE POWER OF DNA
TO WORK IN YOUR LIFE

glenn mcgee

WILLIAM MORROW
An Imprint of HarperCollins*Publishers*

HarperCollins books may be purchased for educational, business, or sales
promotional use. For information please write: Special Markets Department,
HarperCollins Publishers Inc., 10 East 53rd Street, New York, NY 10022.

FIRST EDITION

Designed by Fearn Cutler de Vicq

Printed on acid-free paper

Library of Congress Cataloging-in-Publication Data

McGee, Glenn, 1967–
    Beyond genetics : putting the power of DNA to work in your life /
Glenn McGee.—1st ed.
        p. cm.
    ISBN 0-06-000800-8
    1. Genomics—Social aspects.   2. Genomics—Moral and ethical
aspects.   I. Title.

QH447.M356 2003
306.4'6—dc21   2002044864

03   04   05   06   07   WBC/RRD   10   9   8   7   6   5   4   3   2   1

For Ethan Baker McGee,
Austin James McGee, and
Aidan Bennett McGee—little people running
beautiful but extremely complicated operating systems

contents

contents

genetics again?

as an adopted child, I hated my first biology class. My friends would perk up as the teacher walked us through the biology of 1981, an introduction to the scientific reasons why offspring looked or behaved the way they did. The textbook was great for the boy in the seat in front of me, Brian, who could finally understand why he had big fluffy blond hair, and now had good reason to resent his father for it. Mrs. Tipton used pictures, rules, and diagrams to explain the biology of the family, but I wasn't having any of it. The older kids in the neighborhood set me straight about the facts: adopted kids' families were *biologically fake*.

Every morning at 8:00 A.M.—first period—my mind flew far from the diagrams of peas on the chalkboard. I doodled imaginary machines that would let people control relatedness. My big, noisy contraptions would mix genes, creating links that were not in the textbook. I would combine what was essential about cocoa beans, sugar, and cows. I would use fish genes to swim underwater. I would give turtle genes, for longer life, to my grandparents. I would also transfer adoptive parents' genes into their kids so that I could erase the stigma of being a mystery child; so that I could be part of the "who has Grandfather's eyes" conversation; so that I could blame someone for my inability to catch a baseball.

I failed the first biology test, but I can see now that those imaginary machines were the more important part of class. They let me change the rules in my biology textbook. If my envisioned experiments were outlandish, they raised two perfectly reasonable questions: where do the rules about "natural" things come from? And, why can't we change the rules? I believe it was my great fortune to be brought up in an unusual way at a propitious time.

I learned genetics as an explanatory science about heredity. My children will see genetics in the way I saw my erector set. In one Manhattan museum, ten-year-olds with an hour to spare use detergent and a stick to extract DNA from their hair, then analyze the genes on a machine just slightly more expensive than a laptop computer. The museum has jazzed up its "analyze yourself" booth with ominous, glowing images of the double helix of DNA. But the kids don't care about the big 1980s graphic display. They can dive right into DNA analysis, and if you listen you hear their incredibly imaginative ideas about what they would like to do with "their DNA."

They can change the rules. DNA, genes, genomes, and helixes are yesterday's abstractions. A new generation uses biology as software, as part of a world of virtual and not-so-virtual adventures in a realm once reserved for Nature and God. A new generation sees inheritance not as a big family picture but as a big Internet connection—peer to peer, sharing files, learning a lot and learning it quickly.

For most of the twentieth century, genetics was an obscure kind of biology or an excuse to discriminate against those with disabilities or the wrong kind of skin. In the 1980s, an Apollo-type mission to map the human genome began, sporting astronaut-style superstar scientists, a large federal budget, and a big, ambitious name: the Human Genome Project. Today genetic science has left those early experiments and their government origins far behind.

The mystical promise of genetics has come to fruition, and it is called *genomics,* the systematic study and development of genetic information using tools from the information sciences. Genomics is what happens when genetics is subjected to mathematics on the scale of mapping the interactions between lots of genes and lots of traits; it is *macro-*genetics, and more than that, an operating system. I will refer to the operating systems—rules, procedures, silicon-based hardware, and carbon-based organisms—once analyzed—as Geneware. Geneware is revolutionizing virtually every aspect of plant and animal nature and is on its way to revolutionizing ordinary life for all of us. A new generation is much more comfortable with the "infotech" side—genetic information is the tool that defines opportunity in our century.

Most of us barely understand the social and political significance of the "mapping" of the human genome, much less the so-called "biotech sector" of businesses and their revolutionary new genetic technologies.

Still more of us, especially those whose lives do not involve constant work with computers, do not see that the broad advance of genetic technology is well on its way to rewriting the rules and vocabulary of human life. It is doing so in a way much more dramatic than even computer technology, which transformed the dominant units of human interaction from atoms to bits.

The storm on the horizon is enormous. Suddenly you are an operator of your own DNA—of your own geneware. In the twenty-first century we will change our genes, though much of the miracle will be subtle; we will be altering the environment and transforming nutrition rather than cutting and pasting genes in the bodies of our friends and ourselves. The amount of genetic information at our disposal will be extraordinary, and the combination of silicon-based hardware and genetic software—geneware—will enable radical new choices to be made. If control of the future has always been a central struggle of the human organism, biological efforts to control the operating systems of life will surely turn to attempts to make better children or to enhance those children after they are born so that their genes work to maximum capacity. Let's give my child better-than-average vision. And if my child can have better vision, why not me?

Much of this book is a look forward, but only a few months or years forward, to choices not yet being debated by the inventors or potential users of an entirely new world, one filled with genomic choices. You can download Web information now about what your genes mean or how much exercise you should undertake if you have a particular gene. By 2004 those same websites will upload information from the genetic samples stored in your pocket gene analyzer and return customized information about the relationship between your genes and your current health. No drug, no medical device will appear after 2010 without having been customized for a class of users with similar genetic information. Some of your genes will tell you that you have special advantages, and home genetic technology will let you choose activities and environments that play to your advantages, or challenge you in areas where you have a deficit compared with others in the database. Is your child "gifted"? In a decade, a finger prick and ten minutes' analysis of DNA will yield not only an answer but some suggestions: ". . . eyes suited to visual arts and complex sports . . . not insurable for contact sports due to 40% hereditary risk of trauma-triggered arthritis . . . assertiveness training may be necessary . . ."

My own greatest fear is not that society will go too far, somehow losing our or our children's "humanity," but rather that it will commercialize technologies that do not yet work very well. Genetically engineered tomatoes that taste good, genetic tests for cancer susceptibility that really tell patients more than patients could learn merely by checking out their family history, and gene therapy that is more effective than conventional therapy—all have eluded science despite billions of dollars in government and corporate investment. But this has not stopped the development of massive programs to engineer plants, test patients, and conduct gene therapy.

Just as Nicholas Negroponte's landmark book, *Being Digital,* inspired an entire new way of thinking about the role of computers and digital technology for a *digital* generation, it is my hope that this book will inspire new thinking about the challenges and opportunities that face all of us in this generation that is *genomic.* The genome is here—but not only as a scientific project; it is also a set of tools and technologies that envelop the entire practice of health care and will soon be central to all of our lives. No middle class suburbanite will be without home DNA analysis in 2010. No impoverished African child will be treated by physicians in 2005 without first volunteering a DNA sample to help make developed world genetics cheaper and better for the suburbanites.

We are in the midst of a revolutionary shift from *thinking* about individual genes to *being* genomic, from examining what is possible to changing many of the institutions and rituals that today make us human. The generation that can embrace genomics, though, will enjoy a future in which enormous promise and incredible imagination reshape what it means to be alive and happy.

This century's children embrace the fat and happy cloned sheep, Dolly, as a lovable curiosity and line up by the hundreds to play with a cloned mouse in Chicago and Tokyo. Fear and loathing, the "wisdom of repugnance" of conservative Chicago bioethicist Leon Kass, is *so* five minutes ago, as are the awkward, stultifying hearings of countless commissions convened to discuss the social aspects of genetics and cloning. Traditional politics about abortion, technology, and ecology just do not resonate with new needs, priorities, and values.

As commissions of old white men hold "public hearings" attended by five or six bored drug lobbyists in London or Sydney or Washington, thousands of twenty-something-year-olds watch movies about genetic engineering, or buy a piece of a stem cell company from their PDA, or read 500-word editorials about genomics on the web.

It is time for new rules.

The problem with "expert genetics" is that biotechnology is no longer merely a matter of basic lab science; to understand this fact, though, you have to begin by thinking about where you are willing to go with the power that genetics will bring. It is likely that within three years I can have a portable DNA representation of "me." The digitization of my genes will do for genetics what the digitization of music did for entertainment. I will be able to e-mail my genes, to sell them on eBay, to use them as the basis for art, and to have them analyzed on the fly on my PalmPilot. The potential for portable, wireless, commodified genomic information is staggering, and the ethical implications cry out for discussion in a public forum. Clergy, clinicians, and high school teachers have only begun to play catch-up as this technology comes online. That is what I had on my mind when I set out to write *Beyond Genetics*.

Around the world, geneticists, bioethicists, clergy, and politicians seem to be stuck at a kind of perpetual "starting point": long on rhetoric about the promise of genetics and filled with stories about the heroes of genetic adventure. They come up short, however, on real substance about what the world of genetics has already done to human life and on the pragmatic and ethical implications of genetics for the next decades.

The future of genomics involves the creation of drugs, interventions, and services that extend and improve life for those who live, and preserve and make more accessible the contributions of those who live no more. Better memories, stronger bodies, more self-control, and less difficult growth and adolescence clearly are on the horizon. The more important question is, at what cost?

8

As a bioethicist, I am supposed to help to make sure that inquiry into medicine and science does not leave human values behind. Because the genomics revolution will challenge deeply held values about what it is to be human, no assessment of new technologies is complete without an honest discussion of how humanity's best time-tested values can keep technology working for rather than against the flourishing of all of us, our progeny, and the Earth.

Predictions made by those who have struggled to think about the future of genetics have largely turned out to be false and misleading. The idea that one can simply "map" a single human genome, the fear that there will be genetic discrimination everywhere, the belief that genetic tests lead to gene therapy, and the vision of custom-designed babies all washed out in the 1990s as foolish and clumsy predictions, but only after billions of dollars were spent in their pursuit. But that does not mean prediction is impossible or unwise, merely that not every horse can run in a race.

In this book I will offer a practical guide, "A User's Guide," to the way that computers and genetics, once mixed in our lives, open up new choices. We will upgrade from 1950s models of health, identity, and reproduction to geneware. I will debunk the myths, describe and critique the no-longer-imaginary machines that are here, or practically here, and be honest about what kind of choices you will have to make in the years to come.

Whether you are ready for genomics or not, it is here. It

will wend its way into every corner of your life more quickly and with greater impact than any technology since the miniaturized digital computer. You will be forced to reconcile any fear of "genetic engineering" with a barrage of new information, products, and services from the world of genomics. Genomics will change everything about your world within a few short years, and most of the changes are already underway. What is being created is software and hardware (geneware) to analyze and manipulate DNA for all sorts of purposes.

Policymakers will have to grapple with both the intended and unintended consequences of this geneware—as new technologies and new possibilities emerge at blinding speeds. This book will help you navigate new terrain and make informed decisions about some of the most profound questions you will ever face in life:

As genetic tests multiply and become more accurate, more inexpensive, and more accessible, how will I decide whether or not I want to use them? You are a telephone call away from finding out whether or not you have a lethal gene in every cell of your body. But genetic tests are *risk* tests—they tell you about susceptibility ("predisposition") but do not not offer certainty about outcomes. What does a 74 percent chance of getting prostate cancer mean to you, your family, and your future children?

How much should I tell others about my genes? Should you tell your children, your spouse, your employer, your life insurance company what you learn about your own hereditary risks, or lack of risks? Access to insurance and employment will become key concerns as genetic tests reveal previously unknowable information about your health.

Will you be able to tell the difference between gene therapy that presents a real potential for benefit to you, and gene research that is more likely to harm than help? Most Americans believe that genetic tests lead to genetic therapy, or cures. But most gene therapy is experimental, and it can be deadly. Even harder choices will have to be made about the likely successor to gene therapy, stem cell therapy.

Is genetic food technology a way to improve the quality of life, or a risky sham? Does it make sense to spend billions on genetically engineered foods if they do not bring anything extra to the table?

Should companies be able to patent, buy, and sell your personal genetic code? Companies are already patenting thousands of individual genetic codes they feel will have commercial value in developing treatments for disease. What does it mean to your identity, your religious beliefs, and your rights to privacy if your genes can be bought and sold, transmitted, and used without your consent? Would you sell a gene for a quarter, a dollar, a yen? Are you still going to leave your baby's placenta if you find out that the hospital stands to make millions on genes from discarded tissues?

beyond genetics

Are you ready to plan your future in terms of genetic potential? The information in your genes will have profound implications for the consumer choices you will make in the next decades. The potential for both harm and good from the "digitization" of your genes is staggering. How will you make those plans? During an eight-minute doctor visit? On the Web? With a Magic 8-Ball?

If you're an infertile parent-to-be, do you know what questions to ask about your future, genetic child? Today infertility is often a struggle to get your genes, or at least good genes, into a child. But what makes a child "yours" in a genetic era? What will you trade to have "your" child?

How important will it be for you to make sure your children have a good genome? Parents are now able to expect more and know more of their children before they are even born. Should you be able to use genetic testing to "design" babies? Should you be allowed to refuse a particular type of child based on genetic testing?

Geneware has so far been a blinding, massive rush toward more control over the atoms of heredity, akin to the rush to get movies onto videotape, then onto laser disks, then onto DVD, and eventually into chip-based personal viewers. It is easy to see the advantage of better and more accessible entertainment. It is equally easy to see that instant, constant, high-resolution entertainment has had an impact on human perception and art.

Will the commodification, replication, ownership, and engineering of the genome result in a society that is better and more diverse, or in an impoverished longing for some outdated notion of perfection? Perhaps both. But it has become clear that a new way of talking and thinking about genetics will "give birth" to new public and private institutions that are so different from those of the pregenomic, predigital generation as to make inevitable the loss of much of our parents' reality.

Negroponte commented on the seeming irony of a paper book about digital life. It is true too that an unusual book of any kind about genes—letters about coded letters— is both ironic and dangerous. But this book is, as I acknowledge later, the product of a whole lab of young scholars who supported me not only in getting new information about computers and genes, but also in supporting the book with a special website at *The American Journal of Bioethics,* my journal, published by (appropriately enough) MIT. You can find it at http://bioethics.net/beyondgenetics.

chapter one

bits and genes

humans have been thinking about heredity for as long as they have had to eat. The domestication of animals involved both the elimination of weaker cattle and the use of selective breeding. Herds in Africa were built, as early as recorded history, on what appeared to be the "better born" animals, whether they were sheep, goats, oxen, or camels.

Assyrian records indicate that as early as 5,000 B.C. crops were manipulated one by one through a process that would today be called *artificial fertilization:* the deliberate replacement of the typical activities of sexual reproduction with other activities, often odd in their execution, with the goal of making reproduction more efficient or improving its outcome. For example, the date palm never really had a chance to have normal reproduction, because date palm trees didn't produce delicious dates; people wanted dates that tasted better and came to fruition more quickly.

Before humans had any big theories about genetics, the manipulation of animals and plants became an important part of economic growth. Politicians made choices that changed the world of genetics perhaps as much as tens of thousands of years of evolution did. Some species suddenly got a wide berth—became sacred, or became fashionable as something that humans liked to eat, have as pets, or wear—while other species didn't fit into someone's long-term plans and were destroyed entirely, clear-cut from everything but the fossil records.

Heredity has also always been an "issue" throughout human history. Mostly, the issue has been that people— typically entire cultures—decided that some heritable trait was undesirable and have noticed that this trait might be inherited.

The people who do not inherit whatever it is that is undesirable define what it means to be *born healthy* in the view of the culture, and their health can qualify them for special treatment. The better the family tree, the better the social standing. Few cultures in Western or Eastern history of the past four thousand years have failed to hold up some group as biologically privileged. Only members of the tribe of Levi could inherit the Jewish priesthood. Hindu castes are built entirely around heredity. Most Native American tribes hold or have held that tribal integrity hangs on the restriction of intertribal marriages. To this day earnings and power in the United States can be highly correlated with inherited traits and the stigma or status associated with them. Most of the world now knows of the Raelian religion, which holds that cloning of human beings is the key to eternal life, and that Jesus, a UFO denizen, was himself a clone.

But the power of genetics, from 1600–2000 A.D., has come almost entirely from crude guesses about what will be desirable, resulting in crude changes in human activity, from the transplanting of fish, bushes, and trees (Kudzu, anyone?) to marriage customs and genocide.

The guesses that people make about how genes work can be expressed as formal, even mathematical, propositions. But every guess comes laden with political and environmental implications, and so every fight about how to think about a gene is also a fight about ethics. If a scientist hypothesizes that the inheritance of a gene could make someone smart or criminal or tall or beautiful, it is a safe bet that the debate about how to use that gene will be hotly contested.

Genes can be tools—to make better medicine or to repress the impoverished. They can offer new options or (through DNA fingerprinting) exonerate those who have been falsely accused of wrongdoing. What gives genes their power, then, are the values or ethics used to apply them: the decision made by individuals and society about how genes fit in to the desire to live a good life. Ethics involves making choices, and where genes are concerned, some choices are better than others, depending on where you stand. The question is, who decides how to put value on genes.

What Counts as Family?

Western theories of human heredity were first recorded in the Greek doctrine that asserted that sperm carries hereditary information and "vital heat" from father to offspring. The sperm was thought to direct the form of the baby. Aristotle disputed the notion that females had the vital heat necessary to contribute to the form of the offspring, and also held that traits acquired by parents during their lifetime might be passed to offspring. These early ideas contributed to "big theories" about genetics.

The theory that experiences acquired during life could be passed to offspring helped Greeks account for strange differences in appearance among parents and children. For example, Aristotle postulated that a child whose eye color differed from that of both parents must have acquired the trait from parental experiences. As a big theory of inheritance, Aristotle's was crude but politically effective for persuading parents to be careful before and during pregnancy. As a bonus, the achievements of the great could pass on to their children. Aristotle's big theory did have the disadvantage, as did many others of that time, of being wrong.

The real explosion in the study of the biological family

dates only to 1800. Advancing right alongside is the practical power to effect changes in families. Western populations think of themselves as in control over what a "family" means, and the family is thus the subject of almost all literature since 1850, according to such varied critics as Jean Bethke Elshtain and Cecelia Tichi. We talk about a concept called social and biological *identity* of offspring, a notion forged through years of habitual behaviors by families, courts, and physicians about what counts as a family, what counts as inheritance, and what parts of maturation and development are most important.

Mendel's Peas and Politics

The explosion in the modern investigation of heredity occurred in the early 1800s, when research focused primarily on the problems of inheritance in plants important to a large commercial breeding industry. Scientists sought to uncover laws of biology, applicable to all organisms, that would explain both inheritance and development.

Gregor Mendel's famous experiments with garden peas began a discipline called *genetics*, which concerned itself with the relationship between traits in the parent and traits in the offspring. More interesting than his peas were Mendel's politics. He and those who would follow soon thereafter were committed to the idea that genetics operates like gravity, or perhaps better, like a watch; predictably following laws and rules as inviolable as anything in the universe. How a pea plant grows is merely an operation of the grand laws of nature, not simply a throw of the dice. There was nothing random in Mendel's scientific world; it was all a matter of figuring out what the world was supposed to be like.

How odd and even immoral would today's genetic engineering seem to Mendel? Cloning technology that mixes the DNA of pigs and spinach breaks those basic rules, and Mendel wanted to *find* the natural laws, not break them. He could not have imagined—or likely tolerated—a biology that is sometimes random and subject to fairly fundamental reprogramming.

While Mendel's successors forged ahead in the laboratory, Charles Darwin was conducting zoological investigations and by the 1890's was unraveling the mystery of chromosomes, which contain the basic instructions for hereditary traits. Darwin formulated the principle of *natural selection,* an important step toward the modern account that linked animal and human behaviors to biological heredity. The principle of natural selection dictates that organisms with traits more favorably suited to the environment will reproduce more frequently, and more of the offspring will survive—preserving traits that are conducive to survival in a particular environment.

Darwin's world seems on its face to be much more plastic than Mendel's: it is a world where very dramatic changes do occur, often as the mere result of reproductive preferences. But beneath the radical veneer lurks Darwin's own account of nature as a machine operating according to natural laws.

Stretching the Law

During its classical period, extending roughly from the mid-nineteenth century through the 1950s, genetics was a crude but much vaunted investigation of how organisms inherit the traits that distinguish one creature from another.

But the classical geneticists could not study large populations over time, or identify the millions of bits of genetic information that populations share, or identify the effect of the environment on the genes of plants and animals. Most important, classical plans for improving humanity or plant life had to do with maximizing the operation of what nature already dictates. These were not scientists bent on challenging the nature that they observed, or asking whether, given evolution, it even makes sense to talk about a healthy or natural organism at all, outside of particular environments at particular times.

Those who turned to genetic science in the nineteenth century to fortify their political beliefs found a science that supported the reinforcement of existing power structures. Classical genetics, with its crude notions of inheritability of traits, gave power and license to those who needed a way to conduct primitive experiments that might document the superiority of one race or gender or body type over another.

The political implications of Mendel and Darwin were made obvious by Darwin's cousin Francis Galton, who drew from Darwin's work to promote what he termed *eugenics*, the "science of improving humanity through breeding." Eugenics became a much venerated pseudoscientific template for the application of early genetics to human lifestyle, and the memory of various notorious and irresponsible projects conducted in its name haunt genetic research even today.

Revelations about Nazi, British, and even American experiments in eugenics essentially killed public enthusiasm for research in human genetics from about 1945 through 1950. The politicization of heredity that held up some families or species or cultures as healthy or normal, however, did not dissipate.

Naming the Gene

In 1953 James Watson and Francis Crick got the world interested in genetics again. In a momentous nine-hundred-word article published in *Nature,* they argued that DNA has a structure, the now well-known twisting and intricate double helix. This was of enormous importance because, it was thought, the structure of genetic information would help explain the way cellular heredity functions and replicates. The story of DNA is told as a kind of colossal safecracking epic. Once you can get at the double helix, your trip into the hidden treasures of hereditary "identity" is well under way.

A total of perhaps thirty thousand genes, along with millions more bits of genetic information that do not code for the creation of anything, sometimes misleadingly called "junk" DNA, comprise the nucleus of most of the body's ten trillion cells (red blood cells lack a nucleus). Within the cells, genes are organized in structures called chromosomes. Each human cell contains an arrangement of forty-six chromosomes (twenty-three from each parent, paired together). Altogether, some three billion "bits" comprise the hereditary information, arranged in chromosomes, along the double helix.

But what is a "gene"? Mrs. Tipton taught me that genes contain instructions for the creation of proteins and enzymes that control the metabolism of cells. Within each cell, the gene's "instructions" facilitate the specific sort of metabolism that is appropriate to the cell's function. Though all cells in a given individual have pretty much the same genetic information that was transmitted when the sperm met the egg at conception, there is enormous specialization among cells in the body. As a human embryo progresses from one cell to some two trillion cells at birth, the specialized metabolism of cells converts energy into organ systems and bodily structure.

Where Mrs. Tipton stopped talking, things really get

interesting. All of my cells have all of my genes, but only one or two genes are activated in a particular cell, because the system that allows genes to specialize is so astonishingly complex and fluid. Bone cells become bone cells through enzymatic actions, specified by genetic information. Instead of toenail cells in my liver, or eye cells in my fingers, I have cells suited to the part of the organ and part of the body where they reside.

The discovery of the double helix helped to explain the way genes work, to identify the physical place wherein lies special stuff that governs the relationship between heredity and the metabolic activity of the cell. Knowing that there is a superstructure like the double helix to hold genes together explained how genetic information is passed in the creation of new cells and new embryos. And of course the double helix made a marvelous icon for the mystery of life and heredity, a beautiful and elegant form that snakes around as wildly as the other activities of our lives.

The work of these twentieth-century scientists signaled that the era of crude genetic power had come to an end. Simple relationships between people's appearance and their heredity or between their family and their social standing would no longer withstand the scrutiny of scientific investigation in the majority of the world's universities. Appeals to natural law theory to explain differences among people faded as more complex methods of investigating heredity appeared and as experiments designed to identify the "function" of "ideal specimens" failed one after

another. Only with the peculiar ascension of neoconserva-
tives to political power, under the tutelage of President
George W. Bush in the early twenty-first century, have a
few critics of biotechnology, such as Francis Fukuyama and
Leon Kass, been able to sustain claims that might otherwise be laughed off about what sort of people are "supposed" to exist under laws of nature, or why the public
should be terrified of any effort to tamper with those laws.

The Genetic Sewing Machine

The 1970s set the stage for the new technologies of genetic
information. New procedures were created to see the way
genetic information operated within a cell and to link that
operation to a genetic pattern for the whole organism.

Perhaps the most important new technique was the
splicing of genetic information. The splicing process has
been likened to using a genetic sewing machine. The idea is
to take genetic information from one source, mount it atop
a delivery mechanism, and insert it into another source.

First, DNA is clipped out of the source chromosome.
Second, a *vector*, or delivery mechanism, is constructed of
special DNA that is typically taken from a virus. The delivery mechanism inserts the source DNA into the appropriate
place on the helix of the destination cell, modifying its chromosomes. The modified cell begins to follow the instructions of the new DNA, taking as its purpose the duplication

of a particular enzyme structure. As it duplicates itself, it begins churning out copies of the modified gene.

Think of Willy Wonka's famous chocolate factory, which turned a little boy on one side of a room into bits of floating information, then reconstructed them on the other side. If you could add some extra information to the boy while he is floating about, he might land on the other side with chocolate ears. A more sinister example is found in *The Fly*, in which scientist Seth Brundle becomes "Brundlefly" after a fly lands in Brundle's teleportation device along with him right before he (and the accidental fly) are turned into a genetic pattern to be transported. The outcome in both cases is that DNA is moved from its typical environment into one in which it performs a different role. These are fictional precursors to the development of the earliest forms of what I call "geneware": genes operating in machinery, whether biological or mechanical.

Using so-called restriction enzymes to cut DNA like a chemical razor blade, biologists of the 1970s were able to isolate and remove specific segments of genetic information. When Stanley Cohen and Herbert Boyer successfully implanted chromosomal information from one bacterium into another bacterium, it was an important step toward using the new technologies of genetic information. They had catalyzed the process of genetic engineering, and soon human genetic information would be spliced into nonhuman cells, producing gallons of insulin and other useful compounds.

In the popular perception of science, the production of human insulin by yeast cells was stunning. If an important function of my endocrine system can be reduced to a genetic code and used to reprogram a yeastlike organism, how much further can we take this whole "gene" thing? It was argued by many essayists in the early 1970s that what is true for insulin may indeed be true for every part of the human body—there would be "codes," in terms of a gene or genes, for each human function. More important, though, the carving up of bits of genetic information in this way had opened an entirely new vista on the gene, allowing the discrete analysis of the relationship between genes and the proteins or enzymes that a gene typically makes. The relationship between genes and traits could be studied outside as well as within their ordinary settings. Genes came to be seen as a kind of building blocks, like Lego pieces rather than building blocks in animal bodies; one could stack them and make new designs, then test the designs to see whether they are useful or interesting or merely elegant.

Setting Sail

A vision of the future was taking shape, one that depended on a less myopic understanding of human genetic inheritance, a future in which genetic information would be modified for the purpose of a new kind of medicine.

Human or animal or even plant DNA might be spliced into the DNA of a human being to cure hereditary diseases or improve human traits. Genetic therapy in human embryos could allow physicians to replace some or even most of the DNA in a person before that person comes into the world, curing disease or improving human potential.

Technologies in the 1970s provided a way to splice the genetic information but gave only scant clues as to the details of the connection between genetic information and animal function. A map of the human hereditary matrix was needed. The language of mapmakers—cartography— began to take shape in biology as scientists struggled to cope with the meaning of an organism's collective set of genes.

Some described this map, or more properly the "land" it carved, as a *genome*, ascribing to the whole set of human genes a meaning much greater than the sum of its parts. Others were more pensive about such grand expeditions, arguing for a gene-by-gene examination of human and animal heredity, yet the resources and infrastructure necessary for this massive undertaking were not available. Dozens of facilities were needed, each specializing in a small segment of genetic information. Each facility would need funding, and a structure would have to be created to coordinate and oversee the overall endeavor.

Scientists struggled over how far to go in developing a way to study not the way in which nature's genes work but the way in which changes in them make possible new kinds of people and animals that might be desirable or undesirable to have around the house or the laboratory.

Support began to build for a major government-funded project to study human genetics. The Department of Energy took the lead, with James Watson recruited to oversee medical implications by starting a National Center for Human Genome Research at the National Institutes of Health.

The mapping and sequencing of the human genome was under way. The gene was no longer the mission. Scientists were embarked on what journalist Matt Ridley describes as the greatest scientific expedition of all time, sailing on forty ships in forty different directions to chart the dimensions of heredity, with little more to guide them than computer and biological technologies that each knew was insufficient to the task.

chapter two

floppy genes and
rewritable genomes

within three years of the initiation of the Human Genome Project, many of the "astronauts," the guys with the test tubes, left the building, and with them went the grand theories and big scientific dreams of using genetics to map elegant laws about heredity. The people who took over are mathematicians and computer scientists, experts whose mission is not to make maps of the genetic ocean but instead to swim in the sea of data, feeling about for evidence of risks and benefits of genes, making sense of the map and putting it to use in giant algorithms.

In less than fifty years, genetics has become a digital technology called genomics. The genes that were supposedly glued together so that they could be mapped or read like a novel have become bits in a three-dimensional but floppy whole that goes by the abstract name "genome," and which can be programmed, analyzed, cut, and framed but which in any event are indeed more than the sum of parts.

But like most developments in the digital world, the genome, and methods of studying it, genomics, are still in beta testing, rife with bugs and sudden perilous conflicts with previous ways of thinking about the "software" of bodies and lives.

The Matrix

Most cells contain a complete complement of the genes with which an organism began, and you might call that complete set a genome. But the genes in any particular cell may be mutated by any number of forces, from radiation to a virus. Does a cell with five mutated genes out of thirty to forty thousand still contain the "complete genome"?

It depends.

The pursuit of a map of the human genome was built around the idea that *the* genome that begins an organism is pretty much all there is. But an organism might have millions of different genomes in its many mutated cells. Picking the "right" genome is an activity that makes sense only if you believe that organisms are in some sense better off if the genes that they start out with are all intact in every cell. But that just isn't true, and human modifications of various genomes in particular cells of plants, animals, and humans put the lie to a "natural" genomic state even more clearly.

The genome isn't solid terrain but a matrix of information that is constantly being changed in subtle and not so subtle ways by the environment and technology. The map of the genome produced by the early efforts of geneticists turns out to be useful only for finding the biggest landmarks—one cannot use it to find that quaint restaurant in the Tuscan hills. Vital information about disease, human traits more generally, and the opportunity to alter genes all require a much more complex and three-dimensional ability to see how genes and those who "carry" them interact with their environments. The way we think of our relationship with "data" has to change.

Think of Hollywood representations of virtual reality. In 1982's *Tron*, Disney "digitizes" Jeff Bridges, so that he can interact as one of the bits and bytes on a circuit board. The "core processor" is the enemy in this story, as the supercomputer HAL 9000 was the enemy of the astronauts in Stanley Kubrick's *2001: A Space Odyssey*. But Bridges brings the skills of a "user" to bear in the fight against an overwhelming number of little, evil computer bits and bytes. Of course the human user defeats the computer in the end.

By 1999, *The Matrix* tells an entirely different story of the relationship between humans and computer technology. Keanu Reeves plays Neo, a hacker cog in a giant, boring software company, but *because* he is *so* digitally savvy, he is enlisted to fight against the machines that have taken over the earth. The plot is a hilariously cheesy excuse for some phenomenal *cinema noir,* with people serving as pale,

bald, underwater batteries for a giant network of offen-
sively efficient machines; in the end Neo is fighting for
humanity against computers. But it is not the fact that he is
a "user" that lets him triumph over the computers of the
story. It is the fact that he recognizes that virtual reality—
software that enhances, entertains, models, and highlights
aspects of what we might not otherwise notice—is a liberating force. He one day realizes that he does not need to
fight against the matrix of information that makes up his
world. Instead he moves seamlessly from one world of data
to another, ignoring previous physical laws and making up
new rules so that the *bad* computers cannot keep up. Neo
triumphs not because of his physical strength but because
of his power as a programmer.

The technological message is simple and smart: either
you are able to create new projects and destroy old ones at
high speed, in an intelligent way, or you die.

It is exactly the ability of medicine to make quick
changes, to gut its core institutions and start over in the
presence of new evidence or new social needs, that pres-
ages the revolution in genomics. The geneticists in the late
twentieth century were already clinical professionals operating in a constantly changing and fragile environment.
They had practical reasons to embrace new agendas in
their universities or disciplines, and to learn new skills,
rather than fight against the "evils" of whatever new idea
threatened some fifteen-year-old research scheme.

And because the health care institutions of America

have one foot in the business world, scientists in medical schools are often part of the business world in tangible ways. Twenty-first-century business priorities—energy, flexibility, strategic intelligence, and opportunism—permeate every aspect of the hot areas of clinical science. They are part of the matrix.

The scientists who created the Human Genome Project were part of an Ivy era in genetics, a time when Watson and Crick, tenured for life and with labs far from the hospital, played Isaiah: offered predictions, made claims about ultimate reality, and encouraged young colleagues that a new kind of genetic research was on the way. Watson, perhaps the most outspoken and peculiar scientist since Einstein, was the earliest to predict that genetics would move from the marble halls of nineteenth-century college biology departments into the prefab high-rises of medical schools. But most of Watson's fellow prophets did not live to see it happen, or to shape the implications of the new Disneyland that would unfold there: "clinical genetics," "gene therapy," "bioinformatics," "molecular pathology," and "molecular engineering" inhabit large wings of multibillion-dollar complexes in the world's finest medical centers. Half of the residents of these complexes will never see a patient face-to-face.

Unprepared themselves to build a world where biological theory and scientific practice becomes a computer simulation and a corporate intellectual-property problem, they brought their young colleagues and students to the edge, then set them free to build whatever kind of promised land would best advance the mission to map the human genetic code.

The stage for the coming revolution was set by those who understood that computer power was the key to unlock the power of the genome. Harvard geneticist Walter Gilbert saw the coming transition in genetics when he predicted in his article "Towards a Paradigm Shift in Biology" that the understanding of computer power would be the key to understanding life. While Day Zero, the official start of the Human Genome Project, was October 12, 1990, this too was just staging. The actors had not yet arrived. Nor had Neo yet begun to hack his way into the software of genetics.

The New Generation

Universities had to invent a new kind of geneticist, more akin to programmer than to biologist. In predictions made in the 1980s by Harvard's Walter Gilbert and Caltech's Leroy Hood, the prototype geneticist of the twenty-first century is called a *molecular engineer,* capable of crunching tens of millions of DNA bits as though they were nothing but numbers, engaged in projects rather than lifelong theoretical quests, and spending more time on research and entrepreneurship than on teaching, advising, and university service. In fact, it would be just as likely for the molecular engineer to be an employee of the drug or computer industry as to be a professor.

By the early 1990s the candidates actually arrived by the dozens, most in their mid-twenties, and most trained in the labs of the world's top technical schools. Many institutions did call them molecular engineers and set them to work in massive new genetics buildings, funded by the Genome Project and corporate donations. Those who do the work of genomics typically hang on to the holy title *biologist*, but it is a vestment of a bygone era.

Claiming Genomics

The Genome Project brought an entirely different style and goal to bear on genetics. It started in 1989 with those first ships: forty well-funded labs in the United States and counterparts in many nations, each mapping the genes of one of the twenty-three human chromosome pairs, or working on a cluster of traits or diseases. But by 1999, a single prominent geneticist, such as Eric Lander at the Whitehead Institute in Cambridge, Massachusetts, would study genetic information from one million individual organisms, and in any one month might work on thousands of permutations of a gene.

Another kind of biological entrepreneur is Craig Venter, who left positions at the National Institutes of Health and the gargantuan and controversial Human Genome

Sciences to build a genomic empire that would last more than a decade—a lifetime in genetics—around several key concepts and a profound opportunism that would reshape virtually every aspect of commercial genetics and the mapping of animal and human genomes. Venter set out to create an institution whose methods and goals would be more plastic than those at the government agencies around the world who were racing to map the human genome. By 1999 his new corporate venture, Celera Genomics, would bring together the maker of the best genetic research equipment, PerkinElmer, with Venter's own team of scientists and computer experts.

In June 1998 Venter and colleagues published plans to use PerkinElmer's lightning-fast ABI PRISM 3400 fully automated sequencing machine to speed up the genome mapping effort. He would race the Human Genome Project, and while he would not patent human genes, he would reserve the right to "keep some genetic information for his group's internal use," and more important to license the early use of the genetic information his company found, for a fee, so that others could patent the genetic information for clinical use. He also began to sell the technologies associated with his efforts. Celera Genomics's motto would be "Speed Matters." Competitors and those at the governmental genome effort in the United States were flabbergasted by the competition, and the motto they attributed to Celera Genomics showed it: "cash matters, but ego makes the decisions" read one bumper sticker I found on the door of a Fellow at the National Institute for Human Genome Research in Bethesda, Maryland.

It was a brilliant and controversial plan, and suddenly those who had thought of the Genome Project as a kind of Apollo moon-shot mission, as big public science, began to wonder whether or not there would be a Nike swoosh affixed to genetic testing kits; I tried to imagine what it might have been like for a James Bond villain to race NASA to the moon; for an individual to have put up Sputnik as a challenge to his own country.

The genetic information piled up as quickly as the data from the Human Genome Project efforts. News magazines fell all over themselves at the end of the century to get an interview with Francis Collins, head of the government project, on a motorcycle, or Craig Venter on a yacht, but the science journals were also filled with each team's new discoveries: one innovation after another to pull even or pull ahead in the race to map the human genome. Each team debated the rules of the game. What would count as a map? What would count as original work, and what as derivative of the other team's efforts? For roughly a year the world held its breath as a U.S. president was impeached, but two larger-than-life scientists worked even harder than Kenneth Starr. The race had to end, and with that very president's blessing it did.

The "promised land" in the genomics race was the

announcement by the major labs that each had delivered reports to the most important science journals in which each claimed a map of the entire human genome. Much was hidden in that announcement, such as the bitter fight by Collins and others to keep Celera Genomics's map out of *Science*. And the moment of the announcement was preceded by an astonishing last few months in the desert for the pioneers of genetics, for Moses and his people if you will, during which all of those who had promised to complete their genome maps had to conduct rapid computer analysis of the human genome.

A Big Day for Worms

What did they find? Both Celera Genomics and the public team arrived at a single astonishing conclusion, namely, that there are only about 30,000 human genes in the human genome, far fewer than the 100,000 expected. In its analysis Celera identified 26,588 genes but noted that perhaps another 13,000 may exist. This finding is astonishing not because it shows that assumptions by molecular biologists had been so dramatically erroneous but because, as many geneticists noted, people have only twice as much genetic material as a worm, and fewer genes than some other animals. It is difficult to imagine a more devastating blow to the idea that humans are utterly different from all other creatures.

The human genome may contain only thirty to eighty

thousand genes, but there is an enormous amount of information in the map that places those genes in physical and functional terms. Three billion letters assembled continuously, the human-genome map is impressive not for what it lacks in comparison with the genome of the fruit fly or the mouse but for how unintelligible it is. Here is the biggest book ever written, a transcription of one way of reading the mystery of nature, and one can only reel at how incomprehensible is the genome without the application of highly sophisticated data analysis. Even the computing power assembled for the first map was not up to the task, able only to read small numbers of letters at a time and then only with a high error rate.

Genes and Genomes

Genes are bounded by bookmarks of a sort, called *introns* and *exons*. Across the genome, processes called *transcription* and *activation* facilitate the "turning on" of the DNA between those bookmarks, so that the DNA, which we call a gene, is able to do its job, the creation of proteins or enzymes.

Genes matter. A single aberrant gene can cause many crucial cells to produce a lethal protein, or to fail to produce some enzyme the body needs. Such a "lethal gene," as Francis Collins likes to call it, can kill you. But the human body is infinitely more complex than single-gene diseases and defects would suggest.

Geneticists of the Genome Project began to have the

capacity to investigate illnesses not only in terms of mutations of specific genes but in terms of many genes and
many other relevant factors: where the person lived, what
he or she ate, what other diseases run in his or her family,
what kind of drugs he or she took.

More and more diseases, including many that had previously been understood in terms of a genetic light switch
(on/diseased, off/healthy), suddenly became much more
complex. Diseases could be associated with many genes—
polygenic—and genes could be associated with many diseases, and with healthy or even better-than-healthy traits at
the same time. A gene that causes sickle-cell anemia, for
example, also protects against malaria. And genes that
affect one person in one way might affect another in an
entirely different way.

Most important, genes seem to act in concert with a
whole range of forces and factors; the creature who
"houses" genes lives in an environment, and of course for
each gene and each activity there are tens, hundreds, or
thousands of other genes that also may play a role, and a
host of electrical and chemical interrelationships in the
cells that house the genes, cells that contain more than just
genes.

The idea of a gene "for cancer" is no longer a sensible
explanation of how genes relate to cells that have begun to
replicate out of control. Rather, scientists speak of the roles
of many different parts of the genome and of the environment that cause, or *part-cause* (that is, that play a necessary

but not sufficient role in contracting), a disease. The technology to think about genomes and the complex relationships of genes within environments—rather than genes as simple causes—blossomed as part of the Human Genome Project's rush to develop new ways to think about the relationships between genes and other genes, and between genes and their "hosts"—you and me.

Blue Gene: Computers to Author Geneware

Each group working on mapping the human genome relied throughout the effort, and particularly in the final months, on computer power. The aggregate pool of human genetic data unearthed could not be compiled by the fastest of the then-current generation of supercomputers in one hundred lifetimes, to say nothing of the power that would be required to correlate all of the data with multiple traits. Traits, and the genes that cause or correlate with them, have become bits of data in software programs that selectively examine genetic information and put it to use. Innovation in the relationship between the software of DNA itself and the software used to understand and control it defines genomics. And as humanity becomes better at using computing power in genomics, it gains more power over creation and life itself.

Genomics is more than a new and much more acCelerated version of the classical studies of heredity. It is the new mathematical and programming effort to create operating systems that can organize and read genetic information, comparing genes in any two, ten, or ten million organisms. The effort to map all of the genes in the human genome required cumbersome operating systems and huge computers. But that is where the digital revolution comes in: operating systems get better.

The computer technology that has pressed the world of genetics forward is driven by the most important powers of the first and second generation of computer hardware. Caroline Kovac leads a team at IBM's Life Sciences Unit that has developed the Blue Gene supercomputer, which processes massive quantities of genetic data to model complex operations at the molecular level. The Blue Gene is aimed at identifying how proteins fold, a basic biochemical process that, when gone awry, can lead to disease. The relationships between genetic information and the particular protein product are complex and involve an enormous amount of data. So Blue Gene calculates and sorts that data at a speed approximately four times faster than the published speed of the supercomputers in the U.S. Defense Department's nuclear-response system.

Educated Guesses

One might easily argue that the key to the genetic revolution was the early ability of big machines to make mathematical calculations very quickly, along with the evolution of better machines: large-scale computer networks, databases, high-speed computer technology, and miniaturization. But the intellectual experiments that animated these machines were in fact much more important.

Genetic technology has to do more than tally up the numbers that reduce much of what is essential about an apple or an apple tree to data that can fit on a CD-ROM. It has to make guesses about how that information works, and about how to play that information so that it makes sense.

The sequencing of the genomes of organisms large and small comes as a major achievement, but the meaning of the achievement is still utterly unclear. It would appear that all one needs to do is identify the portion of the genome that is actually genes, translate the genes' coding regions so that one can establish which proteins are encoded, identify other areas of the genome and of other organisms' genomes that might be similar, and predict the function of the gene with models. Big computers would do this very quickly. It sounds simple enough, but like all vestiges of the original plan to map the genome one step at a time, it turns out to be much more complex.

The main problem with computer technology in genetics is the *guess*. It is still unclear what counts as a gene, and scientists cannot predict which genes actually play what role in the ordinary behavior of a particular organism. One definition of a gene relies on the notion that genes correlate with traits, and attempts to map genes onto observable traits. Another settles for much less: a gene can be defined by the proteins that it can cause a cell to manufacture. But some genes do not seem to correlate with traits, and still more genes cannot be defined by a protein.

Computer programmers had long been part of the genetics lab, designing programs to assist in the more rapid completion of known tasks. But now the geneticists themselves had to think about programming, and had to frame their goals in terms of what kind of program would be possible to complete.

By the time David Bowie recorded the ironically titled song "Blue Jean" in 1983, computers and software had advanced to the point that computers could help suggest potential pathways in research. But it was not until the cash infusion of the Genome Project that programmers emerged who were capable of building a machine that could operate in a multidimensional matrix of progeny, proteins, genes, traits, diseases, and epidemiological profiles of populations.

The First Geneware Programmer

Among the more amazing feats of the new generation was the one that saved the public genome project from total embarrassment at the hands of Celera Genomics. Unable to analyze its data quickly and eager to push its effort ahead, the Human Genome Project purchased more than one hundred desktop computers and hired a top programmer. But the project's success was in danger until a graduate student was brought on board who in four weeks wrote a simple assembly program that stunned programmers of an earlier generation. Because he was so fast, the program allowed the public group to gain the lead in the "race" to produce an assembled map. Perhaps there is no more telling moment than that announcement by a twenty-four-year-old graduate student, who had spent the better part of two weeks with ice on his wrists to contain the pain from all the typing, that he had written a computer program, hack-it-together-video-game-style, that could solve a problem that would otherwise have left the Apollo mission of genetics grounded indefinitely, while Craig Venter and his teams of programmers raced past.

The Programming Dilemma: Geneware Is Not Easy to Write

There are two fundamental challenges to those who would understand the human genome, and both can only be answered through the inventive use of digital technology. First, there is a sea of data, virtually unmarked, whose sheer quantity and complexity make the task of understanding it akin to mapping the stars in a small and remote galaxy with an optical telescope. The computer is ill suited to simulating the complex real-life roles of genes in human organisms, because there are just too many variables in most gene-trait relationships to model. And even if that sort of modeling could be mastered, there is the matter of how all of those genes relate to one another, a problem that involves physics as much if not more than biology. Second, the genome turns out to be less fundamental than was guessed. With only thirty thousand or so genes to explain an incredibly complex array of human structures and activities, the attention of those working on genetics has turned to the differences between people (and peoples) and to the products of the genome and how they are made.

Programming must help cut through a giant swath of data, on the one hand, and explain the relatively small number of genes on the other. As these puzzles are solved the future of genomics will come into view. If the roles of genes in the life of organisms can be modeled, what Glaxo-SmithKline pharmaceutical company calls "in silico"—that is, if the relationship between genes and other factors can be simulated in a computer program—it would eliminate or greatly reduce the delay in studying genes the old-fashioned way: *in vitro*. To model a gene-to-environment interaction in a computer could take seconds, and millions of permutations could be checked (what happens if skin with a particular gene is exposed to .5, .6, .7, or .8 seconds of ultraviolet light?). To study the effects of environments on genes in the lab takes much longer. The goal is to figure out, efficiently, how many different people, with specific genetic inheritance, will respond to particular drugs and other stimuli.

The medicine cabinet of 2015 will be filled with designer

drugs that take out specific kinds of infections with great precision. But it will also likely contain drugs made just for you and a few thousand people who, like you, have a particular gene. Not a headache gene, or a flu gene, but a "this person responds well to this drug" gene. You'll update your medicine cabinet the way you update your computer software: responding to new techniques, rubbing out bugs in the operating system, and finding new applications that can make life easier. And the very idea of medicine will have been expanded too, of course, so that instead of bottles of pills, most people with chronic illness will eat foods that have been engineered to do more in terms of fighting and preventing disease—and to do it better than pills. Bottled water, air-conditioning filters, laundry detergent, and carpet all can be designed to prevent adverse reactions triggered by specific genes, or to prevent mutations that were caused by the primitive household and industrial chemical products of the twentieth century. This isn't gazing into a distant future; it is practically here.

At the heart of the animal is, well, the heart. There is no simple substance, no magic cell that determines who or what we are. But the low-resolution map of the human genome, and the work of those who put it together, is evidence of the degree to which human ingenuity has turned aches, disease, and human potential into a field of inquiry, with a few set rules and a mammoth range of possibilities for improvement. The dreams of those who imagined that human race or criminality would turn out to be an inheritable proposition, one that can simply be purged through eliminating a gene, will probably not pan out any more than the brutal dreams of those Nazis who longed to purify the human species. But what will happen is that a highly complex language of amino acids, proteins, and rules of interaction between the body and the environment will open up a range of new choices about what kind of world we want to live in.

More important, that world will take a shape largely dictated by the metaphors of the personal computer: you will pick your operating system, and update your software, and watch the kinds of input you provide, all in terms that can be understood by scientists, in terms of a gigantic new kind of software that replaces what was once a mysterious or spiritual concept of the body. What does it mean to operate your body like a PC? What does it mean to think of your life as guided by fate or God or profound choices, if it can be determined that much of your personality and body and habits are essentially a software program over which you have either less or more control than you might want?

chapter three

learning to program your genes

like many of my colleagues, I carry in my backpack a telephone that collects e-mail, a scanner pen that copies books, a computer that plays movies and stores more information than all of the books in my local library. In a decade I will put in my backpack a card that analyzes genetic information from my, or others', cells. I will use it for dozens of things, from choosing clothing to sports handicapping to helping my physician or dentist or psychologist.

If computer companies and computer technology initially make DNA seem more scary, they will eventually make it more comfortable. The same crowd that brings physicians printouts of Web information about new drugs—as more than 23 percent of patients did in the United States in 2002—will embrace the ability of the computer to analyze and give consumers control of information about genetic risks and benefits. It may even be able to turn your Johnny's DNA into an asset for which companies may be willing to pay.

Much depends on the development of new genomic technology and on information based not just on similarities but on differences between one person's DNA and that of others—not just on what is inherited but on what is modified during life.

Similarity and Difference

The puzzle of similarity and difference among people is one that touches every human being. As an adopted person, I admit to being especially interested. My father and mother looked nothing like me, a fact that only became clear to strangers after I was about five years old. While I had probably heard about being adopted before then, from the age of five onward I had to get used to the way that people would contort their faces as they tried to determine whether I looked more like my mom or my dad.

The puzzle of inherited differences among humans is at the heart of virtually every cultural and religious struggle, of racism and of war. The puzzle of similarity is at the heart of the "melting pot" of Martin Luther King, Jr., and of the political philosophy held by countless others. Physical differences that can be measured may matter in political life, but they should not matter as much as certain shared capacities among the human species, such as what King called the capacity to develop "the content of our character."

Just about everyone has opinions about difference: some are fearful and xenophobic, others are disposed to welcome everyone as a brother or sister. There are countless examples of decisions that members of society make to shape each new generation's views about how much their inherited similarity to others matters, such as teaching children to speak only one language or to recognize themselves as primarily defined by their skin color or the cultures that grow up around skin color.

In the twentieth century we envisioned a map that could identify the traits of human beings through our 100,000 to 200,000 genes. In other words, everyone believed that human beings would be much more complicated at a genetic level than other species on earth, making the business of identifying differences in traits a matter of identifying genes and comparing them from one organism to another. But it turned out there are only 25,000 to 40,000 human genes, and the business of identifying differences among humans can't be accomplished by utilizing only the genes. One has to search all of the genome, including the "junk" regions that make up roughly 90 to 95 percent of the three billion bits of DNA that comprise the genome.

Sorting through the junk, identifying bits of genetic

information throughout the genome as a way of distin-
guishing different "kinds" of people, would be the mammoth task that finally separated the old geneticists from the
genome engineers. The task of working with genetic information using the new and confusing confederation of computer programming and health science that has come to be
called *bioinformatics* is not a matter of analyzing genes more
quickly. It is a problem of analyzing the genetic material
throughout the chromosomes rapidly, and matching it with
genetic material taken from other people in order to identify a shared section of the genome that perhaps causes the
production of an enzyme or protein that is a necessary part
of the trait in which we are interested.

Confused?

If I am looking at a single person or ten people with a
disease, and scanning their shared genetic sequences, I may
find something and I may not. If I can identify some bit of
genetic material—whether or not it is a gene—that is
shared by five thousand or fifty thousand people, and then
work from there to identify proteins and genes shared
among the larger population, I really have something.
What is needed is the digital capacity to identify and study
the bits of DNA shared by populations, rapid-fire.

Different DNA

One of the most basic operations in bioinformatics
involves searching for similarities, or *homologies*, between a

newly sequenced piece of DNA and previously identified DNA segments from various organisms. Finding near matches allows researchers to predict the protein that this DNA makes possible. A popular software program for comparing DNA sequences is called BLAST, and that is exactly what it does—rockets through the available information online and in stored samples to determine where the homologies exist.

The effort to find the differences, rather than similarities, in DNA among a population is equally basic, and increasingly important: drug companies around the world have spent more than two billion dollars on the identification of single nucleotide polymorphisms, or SNPs. SNPs are places in which a person's DNA differs, if only slightly, from most others. They provide markers for everything from a person's genetic propensity to develop a disease to how he might metabolize a drug.

Studies of genetic variation in human populations began with the identification of blood-group frequency in World War I. Identification of certain differences among peoples has long been essential in some kinds of clinical treatments, most notably transplantation and blood transfusion. Diversity among peoples is exceedingly important in new forms of medicine and research, and there turns out to be an enormous amount of it at the genetic level.

The SNP Map Working Group, a massive collection of companies and government groups aimed at finding SNPs, identified 1.4 million of them in just a matter of months, indicating that there are millions—perhaps many millions—of different inherited variations.

SNPs are important clues that people are of much more diverse ancestry than was ever suspected. Understand that human difference is much more complex than was ever dreamt of by those who classified races and tribes over the past one thousand years. Science has begun to head down that path. The absence of SNPs among the majority of genes—the absence of genetic difference between most humans for most genes—suggests as well that there is much that humans share, even where the genes in question were previously thought to distinguish groups from one another. Any so-called "genetic basis" for the classical but long controversial category of "race," distinctions among humans that are observable, typifying, sociological, and genetic, has also been cast into strong doubt by evidence of the lack of similarity among those within so-called races. As interesting as the SNPs themselves have become, their presence is first and foremost a way to study the mammoth pile of otherwise undifferentiated data in the human genome.

If there are many SNPs, there are also millions of clues as to what an SNP might correlate with in any particular patient. For this reason, leading SNP scientists insist that it is not enough to gather anonymous samples of DNA from millions or billions of people. One must know what diseases each person has and has had, what his or her family suffered from, and virtually anything else that is possible to know. Did the guy with this SNP live in the mountains? Near a toxic-waste dump? What was his blood pressure three years after he was diagnosed with heart disease? What is needed, in other words, is a massive collection of *different people's* genetic information, containing SNPs, genetic data, disease information, and hereditary information.

SNPs, Mutations, and the Very Idea of a Genome

Genes in many of our cells are changed all the time by virtually every environment we encounter. It has long been known that genes in particular cells can be modified. The twentieth-century explanation of such change is framed, though, as an isolated event, a mutation or, in the redundant, popular term, "random mutation." Many "mutagens," or substances and forms of energy that seem to correlate highly with mutations in the DNA of cells, have been identified. To see this you have only to look at the skin of a person who has spent a lifetime in the sun, on which millions of mutated cells can be seen with the naked eye in the form of moles and growths. Yet most scientists continue to use language that misleadingly suggests that the vast majority of the cells in a healthy human organism contain the same genetic information that was created when our biological parents' gametes fused during conception. It is a bit like holding on to the idea that everyone who buys a PC in 2004 will be using the same machine in 2007. So many parts are upgraded, break, or are changed by software, that most PCs bear little resemblance to the machines that were initially purchased.

In this century it will become more and more clear that the effects of mutations on cells and organisms are ubiquitous, and that they can be understood and harnessed. Moreover, if a significant percentage of human cells contain mutated DNA, then it no longer makes sense to speak of a human as primarily a product of the genes inherited from her parents. Genes may be a death sentence or may help you live a longer life, but the genes that do so may have been mutated during your last trip to Colorado or with that last meal you ate.

The classical model of genetics assumed that there is a specific set of genes with which most organisms in a particular species are born; it is called "the human genome" because it reflects similarity among the vast majority of relevant genes and because those genes correspond to traits that are natural (or at least typical) in a human being. This is the reason that there are perhaps forty, rather than thousands, of people whose DNA is being sequenced, that is, decoded and added to the library of the Human Genome Project. The mapped genome of an outwardly healthy person would, it is thought, be sufficient to allow those who are sequencing its DNA to understand the relationship between that person's genes and that person's traits, between *genotype* and *phenotype*. While the genome in any particular cell might have some mutations, caused by exposure to mutagens, it was thought that these were (as the word *mutation* denotes) random and slight modifications, not essential parts of the relationship of an organism to its environment.

But genomics and in particular the study of toxicology and of SNPs has opened up evidence that there is more to mutation than meets the eye. Three kinds of things happen to the genes in all of the cells of every organism's body, each of which suggests that while it may be helpful to think of the DNA you have inherited as a genome, it will be even more revealing to identify the DNA you have picked up since your conception.

Geneware Corrupted by Viruses

First, there is what I call *immune hijacking*. When we are exposed to smoke, pesticides, chemical agents of other kinds, and other toxins that our cells recognize as foreign and dangerous, there is the creation of *memory cells*. These cells create a strange form of the genetic material RNA, which acts like a virus and is able to free-float in blood serum as well as in viruses. This RNA can cause cancer, birth defects, and autoimmune disorders. Memory cells and the free-floating RNA bits have been isolated, for example, in the blood and organs of those complaining of Gulf War syndrome, who were exposed to massive amounts of petroleum distillate and other chemicals. The effect of this free-floating RNA is that the body creates a way of reprogramming the DNA in other cells, which results in massive changes in large numbers of cells.

Second, there is *amplification*. This is the phenomenon now well known to biologists of cancer, involving the genetic events that lead to tumor invasion and metastases of cancerous tumors. Mutation of a cell following exposure to a mutagen results in the activation of the mutagenic DNA, and subsequent amplification or, more simply put, rapid growth of cells containing the mutant gene. Cells containing the normal form of the gene are slowly overwhelmed by the new cells, which form a strange new kind of growth in the body, which has come to be called cancer.

Third is what is called *hypermutability*, caused by radiation that affects the entire genome, rather than one gene at a time. Exposure to radiation can cause more genes in more cells to mutate more quickly than would be the case in amplification or immune hijacking. Massive radiation exposure cannot cause the sort of thing fancifully portrayed in the *X-Men* comic books. Most of the time, massive changes in the entire genome of an organism will cause the organism to simply die, as it loses all of its healthy cells to mutation. The progeny of affected cells are often terribly mutated, whether by X rays, alpha particles, or irradiation. This is particularly true where progeny are already at risk, e.g., from injury caused by cloning. The result is not that the patient develops super powers. The effect is that many of the cells in the body are changed and possess slightly or greatly altered genomes.

In many cases, the meaning of these dramatic changes is that you die. But across the lifetime of an ordinary organism, it has become clear that mutations occur so frequently that more subtle effects aggregate until the totality of cells might better be described as a set of competing genomes awash in an organic sea of bacteria and viruses, each with its own DNA. The point of understanding the differences in the genes among any one person's billions of cells is not only to discover new causes of disease but also to take genomics to the next level. If genomes are changed in most cells of the body, one can no more map the genome of a single cell in one's own body than one can map a single genome or forty genomes and describe the result as "the" human genome.

Geneware that "ships" with the human baby encodes complex responses to different possible environments, and as genes change, so too do the responses. You aren't the baby your parents brought into the world, or at least not in the majority. You load new software or forget to update the old and find yourself broken.

The changes start early. Research now suggests that during gestation the torrent of hormones flowing from mother to child, and the ability of the placenta to deliver nutrients, can have a profound effect on the development of the fetus, and later of the baby and adult. The key is developmental perspective, something to which classical molecular biology did not pay much attention. Harvard research that once pointed only to dramatic effects of tobacco smoking on the in-utero fetus now identifies, among other things, the effect of alcohol in the womb environment on the baby's later ability to metabolize fat. These are more subtle changes to the DNA in particular cells of the fetus, and they are not lethal or malignant—they just change you for life, like solar exposure produces a mole.

Existing research methods in genomics, particularly those in the United States, are not geared to integrating analysis of DNA changes with analysis of inherited mutations, or to studying the ways in which certain kinds of "mutations" are not random at all but rather the predictable effect of some element of the environment. How ironic that the Genome Project, which began as an effort to identify the effect of nuclear weapons on the DNA of those harmed by the bombs, should focus so exclusively on the identification of what is inherited rather than what is modified during life.

The development of new genomic technology and even of information based on SNPs and research into the incidence of mutations among diseased patients has already borne some fruit, though: the leading mutation identified with prostate cancer is one that the patient acquires during his life. Soon there will be ways to measure the toxicity of your environment that point to the relationship between what your genes can tolerate and where you live and work. But for this to happen, there have to be many, many people willing to donate genetic material and lifestyle information to a very early research effort that is unlikely to provide them any benefit. Few even understand what they are donating, let alone its value.

The Bank: Time to Withdraw or Deposit?

A number of companies have built technologies to identify SNPs and to use them in diagnosis of diseases. Other companies are building massive libraries of human genetic material obtained either from indigenous populations, like that of Iceland, or from diverse populations long studied, like that of the famous families in Framingham, Massachusetts. Still other companies, like Celera Genomics, are actually doing the identification of SNPs, or finding associations between SNPs and drugs or environments. None of these companies make a great deal of money yet from their efforts, though the potential profits seem almost limitless. In fact, perhaps the single greatest barrier to the research on how to improve and use SNP technology is investor trepidation. Recently burned by the collapse of dot-com stocks, investors are loath to put money into small life sciences companies.

Most investors also know next to nothing about genetics

and are terrified of the "soon we will control the DNA" hype that surrounds it. But it is a silly and mistaken fear: genetic speculation is perhaps the most obvious opportunity for investors to participate in and take ownership of the future of the genome.

The role of the investor and stakeholder in genetics is a critical part of the expansion of genetics into the study of large populations and the differences among those populations. The efforts to develop SNP studies are dependent on biotechnology, and on faith in genetic research. The genome is no longer a big public project, although many of the SNP efforts are still funded by the governments of Western and some Eastern nations. They are instead distributed biotechnology programs, which require the "buy-in" of those research subjects who live on Maple Street, and who would need to donate DNA, or sell DNA, for the project to work.

In the United Kingdom, physicians have begun to recruit patients actively into the public SNP project there, inviting 500,000 people between the ages of forty and seventy to give their genetic information and their health records to a study funded by the British government and UK foundations. But patients aren't gleefully giving up their genetic information or their medical records. What to do?

The banking of genetic information has a short and bizarre history. Like most projects in the history of genetics, it began with fear. Most believed that storage of DNA samples from patients would lead to special risks, such as insurance discrimination and governmental misuse of personal information. Early in the human-genome mapping effort, DNA-banking commissions were formed in most of the nations who worked on mapping. Each asked the questions: should DNA be banked, by whom, where, when, how, and who should pay the cost and reap the benefits of the bank?

There was agreement about only one thing across Europe and the Americas, namely, that DNA banking is a special kind of research, and that universities must take great care to protect privacy and to obtain informed consent from those whose DNA might be banked. It was almost a year before it occurred to those working on these commissions that the genetic information of patients and research subjects could also readily be obtained from stored tissue samples that had been given for other purposes. Most states, for example, keep the little pieces of blotter paper that have been used for decades in heel-stick blood tests of every infant. Those paper pieces contain DNA but were not banked for that purpose.

When I and others pointed out the absurdity of guarding DNA banks with police and fifty-page forms, while scientists use tissue samples for DNA tests at will, it was back to page one. And indeed the matter just grows more complicated, because it must also be determined what will count as a DNA test for purposes of using stored tissue samples. Some argue strongly for privacy and the absolute right of research subjects to know everything about what their material is being used for—even if that means going back to the mother of a baby, or to the grown-up baby, or to the heirs of the baby—for permission. A few even insist that if a DNA test with potential for diagnosis of some disease is performed, the person from whom the sample was obtained must be contacted and notified of the outcome. Others hold that whatever consent was obtained to get the sample in the first place would be enough, regardless of new issues that arise in the examination of the subject's genetic information. But everyone had to ponder how this information would be stored—would patients' records in hospitals include a log entry for the use of stored samples? If so, what would the effect be on the patient's life or health insurance? If not, what is the effect on the patient's right to know and on the patient's health care?

The advent of SNP research made the development of DNA banks essential, and questions about what should and should not be done in the storage and use of genetic information, at best a blurry issue, took a backseat to the debate about how to get enough samples. One effort, the First Genetic Trust Company, formed by the chairman of the SNP Consortium, the organization most responsible for finding SNPs, billed itself as the first "genetic bank." The bank would not hold money but it would make an enormous amount of it, and revolutionize how scientists and clinicians think about the storage of genetic information. It holds samples of people's DNA in secure accounts and gives out the genetic information for medical research and diagnosis only with the account holder's permission.

Drug companies have also tried to create such banks. The Framingham effort failed in part because many of those in the study who were famously willing, decades ago, to allow their health history to be used and published just were not ready to have their genetic information used by an odd, neocorporate consortium. It seemed so much riskier.

Of course this is a microcosm of the shift in personal privacy that has occurred due to the Internet. Tremendous fear of a loss of Internet privacy has dwarfed concerns about privacy in other realms of human existence. Here, too, the ironies are thick: those who think nothing of using a credit card in a restaurant (where a hard copy of one's receipt is left at the cash register for a wide variety of staff to potentially see, then dumped into a trash can marked "paper, not food") will recoil at the idea of entering a name or credit-card number on an encrypted Web page. Why? Because Internet privacy is new and its compromise carries foreboding risks whose dimensions are yet to be carefully studied, and because the Internet is a place with no faces, a zone of strangers, most of whom are young. So too with DNA banks, where strangers think nothing of leaving bits of blood and tissue in the emergency room of a major hospital that is sure to conduct research on it but would not take five hundred dollars to donate DNA to a small research venture protected by high-technology security measures and run by a tiny group of twenty-five-year-olds. The problem is fear of technology. Fear not so much of a dangerous box or a shiny, locked-down biotechnology building as of getting swept into a zone of activities that one might not be able to control.

The Case of Iceland

My own adventures in the puzzling world of gathering DNA from large populations began in 1999 in the tiny nation of Iceland. Several colleagues asked me to join a unique study *of* a unique study. The Icelandic government had passed a special law that enabled the government to grant a license to a private corporation to build and operate a nationwide computerized medical-record system. The law was controversial, in part because it was the product of the efforts of a single company whose corporate base was in the United States, deCODE genetics, a private biotechnology firm that was the only plausible group to complete the project. Beyond the database of medical records (HSD) authorized by the law, deCODE also planned to establish two additional databases, one of genealogical records and the other of tissue samples, which together it refers to as the Genotypes, Genealogy, Health History/ Phenotype database (GGPR).

This was certainly a unique effort, which would put at the disposal of a single company, trading on the Nasdaq in New York, a fairly comprehensive nationwide database of information about a large population. It was unique, too, in that the Icelandic population was described both by many Icelanders and deCODE as not only indigenous but also genetically special, because so little immigration and emigration has occurred over that nation's recorded history. Perhaps the most unique feature of the study is that it makes the assumption that people will want to be participants in the databases; patients must opt out to be excluded, and those who are already dead have no choice in the matter. This in the oldest democracy in the world.

What made our study-of-the-study unique is that unlike all of the other research teams who went to Iceland to look at the situation, we were being funded by a gift to the University of Pennsylvania from deCODE itself, which meant that our group of researchers could venture into the belly of the company, where we were allowed to ask questions that others had not been permitted to ask, and to explore the true dimensions of the problem of creating such a database in this way. At a time when deCODE could ill afford bad publicity (a few months before its IPO, or initial public offering of stock), we were allowed to evaluate the company's ethics, no strings attached. It was daunting, and all the more so because Iceland, while beautiful, is as stark as the face of the moon, a place where at any moment one

of several live volcanoes threatens to swallow up hundreds
of farms and thousands of its unique "mountain sheep,"
who jump up and down the sheer faces of dark cliffs
around the island.

The idea is to get the genetic information together, to
mine the SNPs, to link data about patient health to data
about patient DNA, and to determine how those diseases
and genetic markers have been inherited across the nine-
hundred-year recorded history of Icelandic genealogy.
DeCODE promises to pay all related expenses incurred by
the government, plus about $2.50 per Icelander, plus 6 per-
cent of its annual pretax profits up to $2.50 per Icelander.
The promise adds up to a guarantee that deCODE will
give the Icelandic government roughly double what it now
pays in total health care costs.

In return, deCODE will be allowed to link the HSD to
its proprietary genealogy and genetics databases for a
period of twelve years, with the possibility of renewal. All
data entered into the GGPR databases will be computer-
encrypted by the government's Personal Data Protection
Authority to protect the privacy of those who gave genetic
information. The linking of medical, phenotypic, genealogical, and genetic data contained in these databases will
enable deCODE to sell access to a kind of gigantic Icelandic mine, and to reap profit on its initial investment.
The U.S. stock market reacted favorably to the plan, with
deCODE collecting $173 million from its IPO in July 2000.

The Icelandic government has justified the project by asserting that the distributed computerized medical-record system and centralized HSD will help officials better manage the country's health needs, which may be particularly important in light of recent increases in the cost of health care in Iceland. This would be the only way that Iceland could truly integrate its computer patient records. Icelanders also hope for prestige, and to, as many told us, "lure home" those who, like deCODE's CEO, had been born in Iceland but left for better opportunities in science and medicine elsewhere. DeCODE also had nothing to do with fish or metals, and while Iceland had considered building a giant and offensive smelting plant, many viewed deCODE (at the time) as a panacea; as the "other, better option."

Iceland is the watershed in a paradigm shift in how genetic medicine is practiced, and in genomics. If it works, and even if it doesn't, it represents a transfer of power in "field genetics" akin to the transfer of power in the lab; now companies will participate in, and in large measure control, the progress of genomic science. The deCODE effort is all corporate, from the way security is designed to the purchase of the nation's new health-database computers, to the storing of genetic information, to the press relations and banking effort that keep expectations for the effort high.

If deCODE were to go bankrupt, the people would have no government authority to turn to for the return of their investment of faith, time, and privacy; like so many other companies in the digital world, deCODE is so far selling vaporware, the promise of future products. If something goes terribly wrong and deCODE's stock goes more sour than it already has, the company could be much less stable than a government effort.

On the other hand, deCODE represents Iceland's most prominent role in world commerce in history, and its most important role ever in science. These efforts would never have come into existence without entrepreneurship, and the deCODE group promises to offer not only data for those who want to mine the genes of the Icelanders but also new devices, new programs, and new styles of science—all of which could only have originated with private entrepreneurship.

But what does Iceland really gain? The average Icelander may see no real improvement in health, and is likely to be unimpressed by the personal financial gains that come from deCODE. As my favorite iconoclast, Harvard's Richard Lewontin, noted in *The New York Times*, what appears to be an effort of Icelanders for Icelanders could better be compared to the Viking invasions of Iceland—deCODE is really an American endeavor, a paper corporation in Delaware with only a few Icelanders on its board, one of whom is the CEO, Kari Stefansson.

Icelanders seemed to me nonplussed by this claim. In interview after interview they said that their motivation in supporting the deCODE effort was to advance their nation. A nationalistic people with a long way to go before their role would be on a par with other biotech-hungry nations, Iceland's population, or at least a significant portion of it, seems ready to make the leap of faith, trading risks to personal privacy against profound national prestige. Many also spoke to me in broken but sincere English of a duty to return something to an Icelandic health system that provides some of the highest-quality health—and best lifelong health expectations—in the world.

But it is a leap of faith. No informed consent is obtained from Icelanders for the computerization of existing records and information from future health care visits, centralized data collection, and use of health data in studies that involve the genetics database. While citizens can opt out of the HSD by submitting a form to the Director General of Public Health stating that they wish to have their health care information withheld, we found no opt-out forms in two of three clinics we visited. The opportunity to opt out in totality will also end once data collection begins. Almost twenty thousand Icelanders (about 7 percent of the population) have been able to and have filled out opt-out forms so far.

And even if one were to ask Icelanders to give consent, what would you say? "We think we have enough money to build a big system that hasn't ever been attempted before, and whose output we can't estimate, in order to earn a profit some part of which might return to Iceland—but not most of it"? How could you frame the "other options" for those who might give consent? "Instead of giving us your data, you also have the option to keep it for yourself—or to lobby the government for a national initiative that would return more of the money and more of the benefit to the people and less to American stockholders." These aren't sentences that you ordinarily see in informed-consent forms. The leap of faith in Iceland rings around the world, as Singapore, England, and dozens of other nations take aggressive measures to mine the DNA of their peoples; they take a proactive stance in the corporate war to control health and DNA data, and put so much of patients' private, sensitive information on a secured database to be accessed by thirty-two-year-old pharmaceutical researchers in New Jersey. Plus there is that voodoo that is sometimes confused with science: international economics. A tiny blip of the stock market, let alone the recession of the early twenty-first century, and Iceland is back to smelting and processing cod for Portugal.

In New Jersey

The collective efforts of those in the Human Genome Project, far from Iceland, allow the identification of two thousand bits of genetic information every second, roughly 172 million base pairs of DNA every day. It will not be long before the G16, the sixteen big players in genome mapping in the United States and Western Europe, will finish sequencing their allotted chunks of the genome. Their digitally minded colleagues and friends will then polish that sequence into a "final" form, and a high-resolution map of the human genome will be available. I have a poster of the Celera Genomics map on my office wall at Penn, the university where I teach ambitious future physicians. The poster is a massive eight-foot-by-five-foot display of the different chromosomes as published in *Science*. The polished form, using the same style and font of display, would take up roughly a football field. It sounds exciting. But the G16 aren't thinking about that map much anymore. They are racing to identify products, and that means correlating SNPs with human traits.

The race between public and private in the effort to use SNP data was over before it began. Efforts to complete SNP research "publicly," in the sense of avoiding *any* corporate ties, really don't exist. Those who pressed for the old style of genetics have either moved on to retirement or begun working on the genomes of other organisms. An astonishing new array of scientific developments have begun out of the genomics model of collaboration between companies, governments, and universities, ranging from annotations of the genome through the identification of commonly held DNA strands, to banking of disease information. And everywhere there is the new model of decentralized organizations working on projects that are primarily aimed at building institutions that can put genomics to use.

In Philadelphia and Palo Alto

My laptop is very, very fast. A so-called G4 PowerBook, running at one gigahertz with a gigabyte of built-in RAM and a sixty gigabyte hard disk, it is roughly one thousand times faster and has two hundred times the storage capacity of the laptop I used last year. It gets so hot from all of this activity that its titanium shell at times gives me a kind of sunburn. I never shut it down, or at least not until it crashes. When I am not working, it runs a screen saver called "Folding@home," which connects my computer to Stanford University, where a chemist named Vijay Pande has launched a project to simulate the way the proteins in each cell self-assemble or "fold." Like thousands of others, my computer uses its spare time to run software for Pande, crunching information about tiny bits of the folding process, downloading data through my Internet connection to my computer, then running through an endless series of calculations. My computer returns perhaps half a nanosecond's worth of the folding of one intestinal protein at the end of three or four hours of my "idle time." I am idle but my computer folds the secrets of the human body. That is geneware.

Folding@home is based on a technology created, as best anyone can remember, by the people at SETI, the underfunded Search for Extraterrestrial Intelligence that was the subject of Carl Sagan's book *Contact*. Millions of users allow their computers to scan radio signals during their spare time, uploading the results to SETI as they are completed, so that SETI can do the work of hundreds of supercomputers without buying the supercomputers. The software is written to distribute one big project to thousands of participants. It is a bit like the human genomes running in the cells of your body.

Distributed computing is one key to the elaboration of commercial genomics. You may not be ready to bank your DNA at the First Commercial DNA Bank, and you might think twice before giving the placenta from the birth of your child to the hospital's DNA lab, but many Americans will be willing to respond to a "Calling all PCs" plea for help with crunching genetic information.

Calling All Hackers

It is exactly this sort of network, much more complex than that attempted in Iceland, that will allow Johnny in New Jersey, who can afford a thousand-dollar PC and who loves to surf the Web, to become more and more involved in DNA, whether selling bits of DNA, through DNA chip technology, to companies for research or downloading DNA from his neighborhood or family for analysis.

Being part of the geneware generation means integrating hardware and software with "wetware," devices that make biological material into digital programs. Already digital companies and undergraduate hackers have begun to develop the key means, sophisticated systems that convert genes into multitiered systems of genomic interaction.

Around the world there are now complex scientific programs that effortlessly parse the various relationships between genes and the environments in which they are expressed.

Over the next fifty years, though, Western society will develop an entirely new kind of genomics involving the creation of drugs, interventions, and devices that extend and improve life for the living, and preserve and make more accessible the contributions of those who live no more. The key to these technologies is in the integration of carbon and silicon, of DNA running in animal cells and software running in computers.

Better memories, stronger bodies, more self-control, and less difficult growth and adolescence are probably on the horizon. But the future is not quite upon us; the current release of genetic "software" doesn't work that well. In fact it is downright buggy.

chapter four

bugs in the geneware: genetic testing

one in eight women will get breast or ovarian cancer. That means your mother, sister, or best friend, wife, ex-wife, daughter, colleague, or aunt. It means one of the neighbors. It is close. It means that on the day of a miracle of birth, moms and dads with breast or ovarian cancer in their family tree will hold their breath, at least once, when they gaze into the unknowing and unseeing eyes of their infant.

What is so terrifying is the role of cancer in life, of the invader who allows only moments of denial as it approaches with stealth to touch everything: beauty, identity, love, relatedness, planning. Cancer stalks you from within your core.

In a time when death seems almost conquerable to middle-class Americans, surrounded by unprecedented growth in medicine and science, cancer is the statistical wake-up call that nobody answers until the phone is handed to them. It takes little subtlety to capture its effects, and even scholars who resist the metaphor of cancer as "enemy" finally allow that there could be no scarier form of illness than one that results from the transmogrification of one's own cells.

People run to God, run from God, and run to Mexico when cancer strikes. A week after I began to write this book, my friends Karen and Roger of Texas walked along a pier with me, confiding that "acupuncture and visualization aren't working."

Their plans included all sorts of high- and low-tech interventions. They had enrolled in breast-cancer genetic testing. They agreed to talk with me about it because the experience of finding a gene, or rather a "mutation," that ran in the family had been an important part of their lives. Karen no longer had breasts or ovaries because, when she had already been diagnosed with tumors in one breast, the discovery that she had the BRCA-1 mutation was enough to send her "over the edge" into "self-mutilation," the gamble of removing things so that they might not get any worse. She hadn't thought a lot about the decision; it had just seemed obvious that the tissues must go.

Kicking a rock the size of a softball, Roger cried as he said that the prevention was worse than the cure. His friends asked him quietly, he said, what it was like to make love. I felt out of place when he responded, "We haven't, it just isn't, and the breasts aren't the reason, exactly, we just don't. The erotic, isn't . . . a feeling we have anymore." As preventive medicine, this is a bitter pill indeed, but it will be a pill that many swallow as they grapple with the growing but troublesome commercialization of genetic testing for risks too great to ignore.

In the past few years, every American clinician and tens of millions of patients learned of a genetic test for the so-called breast-cancer genes, BRCA-1 and BRCA-2 mutations associated with breast and ovarian cancer. While many patients enrolled in research protocols in American universities designed to test the association between these mutations and incidences of cancer in their families, virtually every major cancer and genetics organization called for a moratorium on clinical testing of patients for breast-cancer genes until the efficacy of the tests could be established. Most genetic tests for susceptibility to disease have been rushed to market with blinding urgency. No long-term studies, for example, were conducted to see whether or not patients from families with only occasional incidence of breast or ovarian cancer, but who have the BRCA-1 mutation, will be as likely to get cancer as a result of the mutation as were the subjects used in the discovery of the gene, all of whom came from families with many cases of cancer in every generation. To do the kind of research about genetic testing that is required to ensure their predictive value, studies will have to be conducted over many years, and in most cases the patent on the genetic test would have long ago expired by the time a test could be conclusively demonstrated to be accurate. This wouldn't be such a big problem, except that the results of just such a genetic test will lead women to have their perfectly healthy breasts and ovaries removed, or to rethink having children, or in a few cases to commit suicide.

In mid-1996 Mark Skolnick, the scientist who won the race to discover the breast-cancer gene BRCA-1, changed everything with his announcement that while research data was early in the gathering, his company, Myriad Genetics of Utah, would begin breast-cancer genetic testing for any adult patient in the United States whose physician ordered it. Research? No, this was to be offered as medicine; as a reliable test.

Myriad, the spoils of Skolnick's federally funded success, has subsequently become a leader in the corporatization of genetic testing. It immediately announced the construction of a forty-five-thousand-square-foot facility in Utah whose purpose would be to streamline and market the process of genetic testing for BRCA-1. Myriad is in many ways a model both of the kind of "manufacturer" making early twenty-first-century geneware and of the bugs in early technology aimed at bringing genetic power to individuals.

The young salespeople at Myriad can tell quite a story about what it is like to work in a supcroptimistic start-up. One told me, persuasively, that "one day America and the world will look back at the early days of Myriad and see a startling resemblance to the early days of Microsoft." It is an ironic comparison, however naive. Within months of its discovery of the BRCA-1 mutation, Myriad had outadvertised, outnegotiated, and outlitigated all of its competition. It closed down or bought out its competitors at the OncorMed Company and in university breast cancer testing programs and developed whole new styles of patenting and marketing to build a business model for total control over breast cancer genetic diagnostics.

Today Myriad holds the lion's share of production capacity in the growing world of genetic susceptibility testing. The labs it built for BRCA-1 and BRCA-2 testing turn out to be just as useful for testing thousands of genes and even for forensic DNA analysis; Myriad single-handedly identified hundreds of victims, for example, of the September 11 bombing of the World Trade Center in New York. Through its massive, if not yet profitable, campaign, Myriad has become the icon for taking the test to the people. Myriad also epitomizes what I have described as "drive-through genetics," the effort to dramatically reduce the turnaround time between finding a gene and marketing a test. It takes only sixty seconds to get your french fries these days. Soon genetic testing will be almost that fast and almost that easy.

Drive-Through Genetic Testing

Karen read about Myriad in a women's magazine and had her test performed by Myriad through her primary-care physician, who confided to her, "I don't know anything about genetic testing, but I can fill out the forms if you can be patient with me while I try to figure out what all this stuff means." The test came back "positive," as her doctor put it. This really means that she has a better than average (one in nine is average) chance of getting ovarian or breast cancer in her lifetime, and by Myriad's current estimates her odds of "getting cancer" could be as high as 60 percent, but might be much lower, or a little higher. A cancer referral and a surgical consult took two hours, less than one quarter the amount of time recommended by the National Association of Genetic Counselors—not enough time to discuss the implications of a decision about removing a healthy part of one's body. Within two weeks Karen had started down the road to a different kind of body, the onset of menopause at age thirty-seven, and the end of her reproductive life. In the course of removing her breasts and ovaries, surgeons found cancerous growths.

Had she cured the cancer? No, and in fact it was as aggressive as ever, as biopsies taken from her lymph nodes would show a year after the surgery. While we walked on the pier, the waves crashed, and Karen talked about how things in her life were "moving very fast and eroding." It was a perfect metaphor for the dilemma of genetic testing.

Things are changing in genetic testing, and changing fast. Ten years ago it required a medical geneticist and some kind of therapist or counselor. Because everyone feared the uncertain but possibly severe future misuses of

genetic information by insurers, employers, and other institutions, genetic testing required numerous interviews and
was obtained mostly in specialized, tertiary-care hospitals.

The real sea change in genetic testing was the Myriad
move. Myriad bypassed all of the potential "competitors"
at university genetics labs and instead mailed 100,000
brochures advertising its test to the customers it believes
will replace medical geneticists as the primary genetic-
service providers: primary-care physicians like Karen's.

Having bought up all of the gene patents for the breast-
cancer mutations, it was free to charge as high a fee as it
liked: up to three thousand dollars for a test. A high price is
some protection against unnecessary and potentially hazardous testing of people who are not at especially high risk
of having the mutation; not too many curious middle-class
patients with no family history of the disease will ask their
internist for a BRCA-1 test when it costs roughly the same
amount as a year's health-club membership. But the price
will come down as demand becomes an important feature
of the Myriad business plan.

Myriad can claim that the cost saves patients and insurers money. Unpublished studies by Myriad showed in 1996
that the use of the breast-cancer genetic test in large populations would save money, after you factor in the value of
early detection and treatment in the one of five tests that it
expected would be returned positive.

However, Myriad does not announce in its brochures

that it discovered more than one hundred new mutations in its first two hundred commercial tests, each of which it has, or plans to patent. Nor does it explain that one in five of its tests that returned positive was of a patient who already suspected she was at risk because of family histories of breast or ovarian cancer. The incidence of BRCA-1/2 mutations among those with breast or ovarian cancer is as low as 5 to 10 percent, that is, 5 to 10 percent of 9 percent of the female population, which is a very small percentage of those who have breast or ovarian cancer. It is especially small when judged in terms of the cost of the test and the inability of a negative test result to allow better health care. Not all of those who have a mutated form of the BRCA-1/2 gene can be tested, since not every form of aberrance in the gene has been identified. Perhaps more important, it has not been conclusively established why the BRCA gene mutations are important to the onset of cancers of the breast and ovaries. The stakes get higher for insurers and employers as you test more people at around three thousand dollars each. And that means that the cost has to come down for Myriad to make the sale to those insurers and employers whom it thinks should pay for the test. Karen paid out of her own pocket.

It may yet be several years before BRCA-1/2 genetic testing is broadly endorsed for adults, children, and fetuses who do not have a family history of breast or ovarian cancer; the cost has yet to cross into the realm of affordable preventative medicine. Moreover, few internists and family practitioners are likely to be comfortable recommending drastic surgery—prophylactic mastectomies and oopherectomies—for healthy women whose genes suggest possible cancer, until more data about the efficacy of such procedures in preventing future tumors is compiled. The data about preventing the onset or spread of cancer in those with BRCA-1/2 mutations through medications, particularly Tamoxifen and Taxol, is also preliminary.

The breast- and ovarian-cancer situation is far from resolved. However, what you can know about the commercialization of genetic testing, and what we have learned in particular from the Myriad case, is so troubling that virtually every commentator in genetics and public policy has called for rigorous regulation of genetic testing that would extend into much more public oversight of the actual labs that offer diagnostic tests, as well as regulations that would govern the claims that are made by these labs and their accountability when, as often happens, they make mistakes.

High-Speed Genetic Planning

Drive-through genetic testing is a phenomenon that has spread far beyond breast- and ovarian-cancer testing; many genetic tests are now offered direct to physicians, and direct-to-consumer genetic testing is available at a small number of "health enhancement" clinics in many nations

and on the Internet. It was a small jump from genetic testing by a physician untrained in genetics to direct-to-
consumer testing. Consumers already spend millions on
in-home HIV testing to avoid sharing information about
their health with insurance companies. Given the issues
Karen faces—planning for her family's life after she dies,
planning her own final years, and putting her relationships
in order—it is easy to see the financial and emotional
temptation that she would have to order a genetic test from
the back of a magazine.

Her physicians scarcely helped her anyway, she says,
and the appeal of a test on the Internet is "that there is
more information there. My physicians have two kinds of
information: drug pamphlets, and magazines filled with
drug pamphlets." Karen isn't in the minority. Studies have
demonstrated not only that clinicians who have little
knowledge of genetics misinterpret genetic tests, but also
that there isn't enough genetic education or money to bring
Karen's physician up to speed on breast-cancer genetic
testing, let alone cancer-genetics testing in general, any
time soon. Society is well along the road to commercialization of genetic testing, with more tests for more people in
more contexts, while at the same time the profits from testing are being taken by investors rather than returned to
public or clinician education or research.

What are the problems facing people like Karen who
want to use genetic testing today? The first is the assump-

tion on the part of just about every group that controls genetic testing that genetic services can be provided without highly trained counselors who understand genetics and the social issues involved in genetic choices. Around the world, studies show that the average practicing internist has had less than one semester of molecular genetics, and only in the past ten years have medical schools begun to put a real emphasis on teaching doctors genetics. Nursing schools are doing a little bit better, but not much and with still further to catch up.

Patients who come to general internists or nurse practitioners for breast-cancer susceptibility testing need comprehensive counseling about the medical, social, and economic implications of the test and a detailed description of genetic probability. Drive-through genetic services just will not work in a population that doesn't know much about genetics, and whose clergy, politicians, and judges know only a little bit more. Karen might be more comfortable with a website than with her current clinical arrangements, but even she says, "It makes no sense to me that I can't go somewhere and talk with someone about this more... slowly."

The second problem is how to regulate the "roll-out" of genetic tests, or rather the development of standards that would show that a test is no longer being studied to prove that it works and is now ready for patients. At present, the clinical claims about effectiveness of tests are totally unregulated. Or, rather, they are regulated by the marketplace: If people buy a test, the company can do more research and accumulate more samples, and, it is hoped, make a more sensitive and predictive test. The product gets better if the public buys more of it. This might be a fine way to research fast food, but it is no way to execute the early clinical development of medical tests. Moreover, what you do not know about the efficacy of a genetic test should not be hidden from you, especially if there are no "second opinions" to be had.

Loading the Program

So how do you make the decision to have a genetic test? How much information is enough? The first step is to understand as much about genetics—about how the genes in your body work—as you can before you start down the path to testing for yourself or your relatives.

The second step is to invest in enough genetic counseling to make an informed decision. Until they say otherwise, assume that your physician or nurse practitioner doesn't know a whole lot more about genetics than you do. As of 2003, there are fourteen hundred medical geneticists and genetic counselors in the entire world. That means that millions of people do not have one within fifty miles. In many U.S. states genetic counselors travel to hold clinics in churches and hospitals. If you live in a rural area, find out from your local hospital's obstetrics department what kind of genetic counseling they use, and start there—even if your problem has to do with a genetic test for liver cancer. Once you find a genetic counselor, listen carefully before you shell out the money required to have a test. Of all the investments you can make, none is likely to pay off more than genetic counseling—a study of your family history, for example, often identifies more risk information than any actual molecular test ever could.

The third step, though, is to go beyond genetic counseling and find out more on the Internet and from physicians and nurse practitioners who treat those who have the condition you are concerned about. Find out what the ten-year morbidity and mortality is for a given disease before you do something draconian based on a genetic test. Does it really make sense to spend thirty thousand dollars to conduct genetic testing on your embryos for the Alzheimer's gene in hopes of preventing the birth of a child with that gene if Alzheimer's will likely be cured within fifty years?

And think in terms of risk abatement versus stress: do you really want to know everything about all of your risks? The tens of thousands who today spend eight hundred dollars on "preventative" MRI scans of their entire bodies often find themselves unable to sleep for months, because they have seen that little growth that almost all of us have—some patch of fat or bit of extra bone that might or might not portend disease. Do you really want to give up your peace of mind—and perhaps any future life insurance as well?

Making an informed choice about genetic testing begins with a recognition: these are the dark ages of genetic testing, and any choice you make today about a genetic test will have unanticipated consequences and might very well be better put off until the technology matures. But when the technology is better, when the bugs are out of the system, tough choices still remain.

The problems with today's genetic testing are bigger than Karen's genes or the test she took: a buggy and creaking operating system is at the heart of today's genetic-testing technology, one that does not have the capacity to give good and accurate predictions that take environment, age, or other factors into account. It is a problem that begins with bad information and involves poorly created genetic tests, irrational ideas about how to balance risks, misdirected fears of genetic discrimination, and excessive hopes for gene therapy. And, unfortunately, even when you can find someone without training to help you make the decisions, their help can be more confusing than no help at all.

What Counts as Genetic Counseling?

Genetic testing is everywhere, and more is coming. But counseling about genetic tests and their implications cannot be accomplished in the thirteen minutes most patients around the world are allowed in a typical visit with their primary-care clinician. If clinicians cannot provide counseling in primary care, how is it to happen?

Hundreds of thousands of tests that can be described as "genetic" are performed every year. Under the old model, ten years ago, it wasn't difficult to find a genetic counselor, and the tests were much simpler to explain. Today, most payer organizations in the United States do not provide many genetic tests, let alone referral to genetic counseling. Around the world genetic testing remains difficult to obtain unless the patient has deep pockets. And genetic counselors, quite clearly the professionals best equipped to ensure full-service rather than drive-through genetics, are at a crossroads, vulnerable to the winds of change in genetic testing.

Counselors are facing a potential increase in genetic testing that calls for the conscription of thousands and thousands of seventeen-year-old biology students into genetic-counseling boot camps. Genetic testing, primarily in large commercial labs, is about to explode, while genetic-counseling is growing quite slowly. And the issue is one not only of bodies but of cost. The rapid technology transfer of genetic tests from small studies run by individual university investigators into large-scale commercial endeavors is unimpeded by regulations or even public outcry. The start-up life sciences companies that run testing today need to cut the cost per test so that it will be possible to build, and to stay in business during the critical years when their test is protected by patents. And these labs are in no position to offer one-on-one counseling, since most are miles from the site where the test is given, and since counseling involves considerable responsibility to provide good and accurate information, and thus entails legal liability.

Billboard Genetics

At least one commercial genetic testing firm has responded to this challenge by offering color flyers and short courses to primary-care physicians in lieu of genetic counseling. I myself have given perhaps two dozen "here are the ethics of genetic testing" lectures to large groups of physicians under the sponsorship of "big" genetic-testing labs like that one, Myriad. Virtually all commercial genetic-testing labs have careful informed-consent documents that, when signed by patients, tend to shift the responsibility for counseling away from the company and to the physician and patient (at least in the mind of the person signing the form, although under U.S. law one cannot sign away one's rights in such a document). The goal is to shift the perceptions of the naive about cost and risk.

Counseling, genetic counselors correctly argue, is more than just providing a pamphlet. The key to the survival and growth of genetic counseling is to build a larger population who can help in the educational effort; to concede that the enterprise of genetic counseling is due for an overhaul that allows others to help. The tendency in the genetic-counseling community has been to cry "Alamo"; to retrench against any efforts to change genetic counseling. There are retaliations against any effort by clinicians or companies to expand counseling to include those who do not have formal graduate degrees in the field.

But if genetic counseling continues down the current track, and the health economy and genetic testing continue to advance in anything like the present format, there will not be many (human) genetic counselors in fifteen years. That is a terrible possibility, and it is essential that counselors avert it by demonstrating the utility of the *counseling* part of their jobs: helping patients assess risk, but also empathizing; encouraging or discouraging patients just like any other health practitioner so that testing is offered and conducted only when it makes good clinical sense.

The problem is the insistence by genetic counselors that they must employ what they call (borrowing from a school of thought in clinical psychology) "nondirectiveness" in all their encounters with patients. Genetic counseling is saddled with the sorry history of eugenics; the early twentieth century was a terrible story of nations sterilizing and even killing their people in the name of better breeding. Counselors do their best to define themselves and their role in strict opposition to any systematic attempt to use public-health in those terrible ways. Everyone knows about eugenics, whether by name or not. The early-twentieth-century effort to "improve" the public's genetic inheritance through British, American, and, most notably, Nazi laws about marriage and sterilization, eugenics was taught as scientific fact in the finest universities of the Americas and Europe until 1945 and led to the sterilization in the 1920s and 1930s of more than 20,000 people in the United States and 250,000 in Nazi Germany. Harvard offered a major in eugenics.

Genetic counselors are terrified, pretty much to a person, that they might somehow accidentally promote eugenics. Their reaction is in some ways more radical than eugenics: every counselor is given the impossible charge to hide his or her own feelings about any prospective test. Counselors have become devoted to the idea that they are supposed to do little more than carefully dispense information.

The fear of influencing patients misses the fact that the effect of genetic testing, such as that for Down's syndrome, is already eugenic in the sense that it casts the test as an appropriate medical technology rather than a "cosmetic" or "elective" procedure. By contrast, if there were no test for Down's available, because, let's say, of a ban on the discriminatory use of genetic information, this too would be eugenic in the sense that it would elevate the life of the Down's patient above that of other potential lives that might be ended through elective abortions. And, if you cannot afford to get a test, or if your insurance gives you a test but no counseling, yes, that is eugenic too. In all cases there is an institutional bias about what sort of people there should be. Whichever tests are allowed, and whether or not patients are influenced in their choices, there is an effect on who is born. Counselors cannot strive for an objectivity, or neutrality, that is logically impossible and socially ridiculous.

There are positions everywhere. The life insurance market is driven by actuarial tables in which risk determines rates, so that if you want to protect your family with life insurance, you will be strongly encouraged to engage in activities that allow a longer and healthier life. The health insurance market is a gigantic mess driven, in the United States, by for-profit allocation schemes and leaves in its wake a gap of forty million uninsured people. If you do not have private health insurance, your ability to prevent birth defects through prenatal decision making and health care is radically limited in the United States and in many other nations. Is that not pressure resulting in change in reproductive patterns? Most U.S. states refuse siblings and first cousins the right to marry. A few governments have skirted charges of eugenics by changing the names of their laws: Japanese rules recently changed from *eugenic* laws to laws about *maternity*, without altering their content. The United States has policy that makes it more likely that particular groups will not reproduce at particular times, such as chemical short-term birth control provided to the unemployed in many U.S. cities at no charge. Eugenics is not avoidable, if what we mean by eugenics is social policy that has an effect on who is born and in what health. Those genetic counselors who doubt this fact should consider that "nondirective" genetic-testing services are provided on the basis of ability to pay. No cash? No referral for counseling. The fact that a few people receive free genetic counseling pales against the fact that many genetic tests, 80% or more, get no counseling at all.

The ideal of counseling without bias has led to an obvious lack of published discussion among scholars in the discipline of genetic counseling about when "direction" is appropriate in testing; when, for example, a testing professional should refuse to provide a test like Karen's for breast cancer on the grounds that the test is not yet good enough, and when it might be appropriate to pressure a patient to have a genetic test in just the way you might pressure a patient to have a test for a sexually transmitted disease or a heart problem.

The way to teach genetic counselors to abandon the idea of nondirectiveness is to teach them to think about the ambiguity and uncertainty of new genomics, and to give them ethics resources that go far beyond "don't push the patient around." What this means for you is choose a good genetic counselor if you are even thinking about a genetic test—and even then spend $100 on a second opinion and three hours on the Web, just to be sure.

Computer Counseling

Whether genetic counseling changes as a profession or not, the digital revolution in Internet education has begun to merge clinical information about genetics with clinical service. I was asked to review the first CD-ROM genetic-counseling tool for the *Journal of the American Medical Association,* a program to counsel for breast- and ovarian-cancer genetic testing. It looked good, but could a computer really counsel patients for genetic testing? I asked students from my bioethics course, and pregnant women who were subjects in one of my studies of genetics in reproductive decision-making, to try it out.

The first two parts of the program were a comprehensive, straightforward tutorial on genetics and breast cancer. Here were movies about all aspects of breast-cancer etiology, risk, symptoms, and outcomes. In several years of listening to presentations by genetic counselors and of watching presentations of genetic testing at professional meetings and community discussions, I have yet to see a better presentation of material on these issues. Both the students and research subjects confirmed my impression.

The third section of the program described the genetic-testing process for the BRCA-1/2 breast-cancer mutations. Here, despite its protests to the contrary in its numerous "I am not a person" disclosures, the software was really competing against the counselor in earnest. And it did well. Pros and cons of testing were explored in detail, including a discussion of confidentiality, insurance and employer discrimination, and psychosocial issues, and the whole program was designed to be updated in a way that we "humans" haven't quite been able to master.

But the program was no psychologist, and it showed in the interactions it attempts. For example, at one point the user was asked to type how she would feel if a test returned a positive result. More than forty participants typed in answers, each receiving the same reply: "These feelings are important and valid." That's right, if you tell the computer you want to jump off a cliff or rob a bank, you get this reply. Most who used the program in my classes found even this interaction reassuring, and pointed to it as an advantage of the programming. But that just reinforces the earlier point that counseling is important in part because the counselor is a *person* with values, capable of empathy. A computer can only pretend to be empathetic. So the trade-off is clear. Better information with the appearance of empathy, or a counselor trained to offer value-free information and an only slightly less limited form of empathy.

The creators of the program, prominent and smart bioethicists Norman Fost and Michael Green, confirm that in their own studies those who have completed the program report higher comprehension of all aspects of BRCA-1/2 testing than do those who have been counseled by a genetic counselor. The CD-ROM is available from the University of Wisconsin and is a very useful tool that is highly recommended for those with fears about hereditary breast or ovarian cancer; unfortunately, there are not yet any widely published software products for other risk-related genetic tests. But the program was 2000 technology. Your PalmPilot will do better, for 20 tests at once, within a year.

Should physicians supplant expensive genetic counseling with software? How much value should physicians and payer organizations assign to interaction with a genetic counselor, if tools such as this one could secure better patient comprehension? Unless genetic counseling reinvents itself, that question is likely to be answered by simple economics. The future of counseling looks digital to me. There remains only one big question: should anyone offer you advice?

What Counts as a Big Risk?

There have to be standards for when it is appropriate to advise someone about their genes—and, even more important, how to present the opinions of those who have thought long and hard about the tests at hand. But is it as simple as letting the standards shape themselves in the marketplace? What about the marketplaces of *ideas*?

The first step is to build forums for discussion of genetic risk, bringing health care workers, scientists, clergy, social workers, public-health professionals, health economists, and scholars together to develop standards for the evaluation of genetic tests- -standards that are neither "bought" by industry sponsorship nor unintelligible to the average physician. The key to these forums is putting genes in context. How do genetic risks differ from other kinds of risk in the human environment.

In so-called "public health" there are a range of sanctions and incentives. U.S. law requires that seat belts be worn because of the degree to which they reduce the risk of injury in an accident. U.S. law requires that manufacturers of cars install air bags. States require vaccination of children, testing of prospective married couples, fluoridation of water. Each rule reflects one aspect of public attitudes toward risk management by the government. Some activities to reduce risk are more subtle. Magazines encourage vitamins for pregnant women without offering the women an incentive to participate. Websites have begun to open public portals on genetic information, translating what only scientists could read a few years ago into information about risk that anyone can understand. Before you can act on risk information as a patient, family member, or citizen, though, you have to be presented with a way to *frame* the risk; something to compare it to.

Some kinds of hereditary risks might correspond to the seat belt: some nations or states might want to legislate that at-risk families who have had multiple Tay-Sachs children must have genetic testing. Some kinds of risks might correspond to the air-bag protocol: nations or states might want to ensure that a set of genetic tests is given to newborns for conditions like hemachromatosis, the most common hereditary disease, a lifelong condition involving an excessive load of iron in the blood. Some kinds of genetic risks might correspond to vaccination policy: a nation might require that parents whose family history includes a genetic risk of particular diseases agree to preimplantation genetic screening before society would pay for *in vitro* fertilization for that couple, or, less dramatically, might insist that parents whose children are at risk for a genetic disease ought to pay for genetic testing of their fetus or themselves to enable everyone involved to prepare financially for the child. Perhaps some genetic tests will seem more like the fluoridation of drinking water, such as future genetic tests for heart disease that make clear that everyone, or most all of us anyway, should drink red wine and take baby aspirin as a preventative measure against heart disease.

The problem with labeling genetic risks, and with identifying the risks associated with taking a test, is that genetic testing is changing, and changing fast. Karen's test is very different from the classic test for Down's syndrome, because while her test does indeed identify a mutation, unlike the Down's syndrome test, that mutation is linked only to a small percentage of the population with the relevant disease. Of those who have the gene mutation that Karen has, 40 to 60 percent will not have breast or ovarian cancer. There are no tests for diseases like the one Karen faces that provide a simple "yes" or "no," and knowing that there is a genetic risk can thus be riskier than not knowing because it opens up such dramatic changes in self-image, and often terrible new choices that may be premature.

New genetic tests present different challenges to law, policy, and personal integrity, both because they test for conditions that are not necessarily hereditary, and because they have, to date, typically been inaccurate. How can you make decisions about genetic tests when so much ambiguity exists about how they are used and interpreted?

How do you make sense of the new variety of hereditary genetic test, which offers you information about a risk rather than a yes or no answer? What does a 74 percent chance of prostate cancer mean for you, for your future children, for your family, for your estate? Would you like to know more or less about such risks?

What *Is* Genetic?

In part the difficulty of answering such questions stems from confusion about what is meant by "genetic risk." The nineteenth- and twentieth-century biologists who saw genes as the "atoms" of human heredity thought that genes *caused* diseases, and indeed there is an announcement of a "gene for" some disease every day. Most disease, however, is not a simple matter of inheriting a genetic mutation that causes its onset.

The more complex diseases offer the best way to see the problem with the classical approach to genetics and disease. Take, for example, attempts to identify a genetic *tendency* or *predisposition* to obesity or alcoholism. If scientists can identify a genetic mutation that is correlated with obesity, and then determine that the gene causes the liver to prevent speedy metabolism of certain fatty acids, it will be tempting to say that there is a "fat gene," a gene that explains why its owners are obese.

But in the twenty-first century it has become clear that such studies do not identify a fat gene. A person with this gene who eats fewer fatty acids than is average in the culture may not become obese. More important, the very definition of obesity, let alone the way in which it is judged, changes monthly as authorities make new judgments about what sort of body is most happy, healthy, and wise, and what sort is least open to specific risks (risks that are themselves being redefined on an almost yearly basis). More important, a gene isn't a *predisposition* to anything. The gene is either turned on or turned off. It doesn't cajole, waver, or worry. It doesn't cause the organism to seek the fatty acid that is in the diet of its society. It only causes certain fats to be broken down more slowly. The critical feature of obesity is a social decision about what kind of risks are tolerable and about what kind of body is healthy. So one would hope that a genetic test for obesity would itself be thoroughly tested in a wide range of populations and a wide variety of possible diets before results could be made available. But if that work isn't done before a test is offered, it does not make sense to call the resulting test an "obesity test" or the gene a "fat gene." No matter how you frame it, an obesity genetic test is no bargain unless it is tied to specific treatments for specific people, and removed from the world of genetic discrimination.

With other genes the matter is not so complex, even though the disease may be hard to define. Genetic tests for a predisposition to heart disease are advancing very rapidly, and within a matter of a few years it will be possible to detect hereditary susceptibilities to what we today call a "heart attack" literally years in advance. Knowing what to eat to prevent the kind of high blood pressure that runs in your family will also be simple, at least for millions whose genetic profile has been correlated with both symptoms and treatment. A genetic test that could help you identify foods that would prevent heart disease would not be a matter of much controversy.

What Makes a *Good* Genetic Test?

The difficult thing is in telling the good tests from the bad ones. One would hope for a regulatory structure that offered testing of the tests, of the labs, and of the marketing materials. But to say that the pace of virtually unregulated genetic testing and the research that supports it, most of which is now conducted in the private sector, has outstripped the wisdom of conventional counseling and testing approaches is an understatement.

Genetic tests are now primarily made available to the insured wealthy, and information about tests advertised to the still more monied few. Myriad began offering genetic tests, without counseling, by mail, to any physician in the world, before any reasonable conclusions could be drawn about how effective the test is in diagnosing breast cancer in families with a limited history. In what was supposed to be a display of ethical power, it pronounced that it would not "allow" prenatal BRCA-1/2 testing. Still other companies, including one of the competitors swallowed by Myriad, had enrolled paying patients as "subjects" in "studies" of the efficacy of their test, using for-profit Institutional Review [Ethics] Boards to maintain the appearance that a careful study was preceding corporate marketing of the eventual, real product. But Myriad killed those "studies" in favor of its ready-for-market approach to developing the test.

Technical complexity only adds to the corporate morass; the number of mutations and diseases related to multiple mutations, or multiple diseases related to the same mutation, has ballooned to the point that now several populations have been tested for one condition but may not realize that the test is predictive for another risk unrelated to the disease they fear and for which they were tested.

The physicians who obtain most of their information about new medications from a drug-company representative (as do the majority, recent studies show) are the same physicians who learn of risks and benefits of BRCA-1 testing from a Myriad sales representative. And, not surprisingly, the treatment patterns for women who have received the breast-cancer test mirror the recommendations of its producers: radical bilateral prophylactic mastectomies and oopherectomies to prevent breast and ovarian cancer, even though evidence about the efficacy of such procedures is paltry and the risks of early hysterectomy may include increased incidence of other forms of cancer. It is especially strong medicine given that the efficacy of single gene-mutation tests in general populations is not demonstrated for any complex disease. None of these tests have yet been demonstrated to be substantially more predictive, or any more cost-effective, than taking an oral family history.

Tapping the Brakes

The question is how best to slow and regulate the clinical introduction of tests. To leave genetics to the market makes no sense. The market did not pay for the initial research; rather it was financed almost entirely by the trusting public. Yet, while on the one hand the public has financed the development of an effort aimed at producing a public map of the human genome, on the other hand it wants to see results from that effort, and further is literally invested (most mutual funds now include some investment in the life sciences) in the success of the companies that do commercial genomics.

The identification of risks will improve as a direct result of technology used to map and identify genetic differences among peoples around the world. Technology that obtains and analyzes samples will quickly become technology that can be used to build databases to allow individuals to measure their genetic information against that of many others with all sorts of similar or different circumstances. The

implications of this are profound and in many ways positive. One danger, though, is that the language of risk itself
has not been clarified. When those who want genetic testing or genetic education seek out information about risks
and how they work, they find fewer than two thousand genetic counselors, and those who are practicing are committed to an ideology that makes it impossible to offer helpful
advice about how to frame risk or make decisions. The fear
of eugenics may indeed *cause* a new "credit-card eugenics"
in which careful evaluation of risk is a matter only for the
individuals who can afford to access the genetic databases
or who can afford to use the technologies that come from
them. That is a problem that becomes more acute as genetic information becomes integrated into the evaluations
that others make of us and of our potential.

What Counts as Discrimination?

Genomics blurs the line between illness and disease,
between feeling ill and having a diagnosis, to the point
where often no line exists. When a physician tells you that
you have a genetic abnormality, it may or may not have
anything to do with how you feel, and may in fact make
you feel bad in and of itself. Genes that have been correlated with Alzheimer's disease, with Huntington's chorea,
with prostate cancer or herpes or schizophrenia are genetic
risk factors, but patients would not necessarily "have" the

disease, or know they "have" or will have the disease, at the time of testing. A hereditary genetic test offers information about risks of senility that were set into motion at the moment of conception. A genetic test for a hereditary mutation is not an ordinary medical test, not only because it will change the outlook of the now-future-patient, but also because it will affect that patient's options. The person with a "disease gene" is now in some sense condemned to be seen as a patient who already has a disease, and as such will in many cases be unable to access the full range of society's services, such as life insurance.

Even if a genetic test turns up a disease that is fully florid, the diagnosis through genes is a difficult one, both because genetic tests seem to many to be truthful and damning, and because they provide information that is not vague in the *same way* other clinical tests are vague. A patient with a biopsy that shows cancer may interpret the cancer, and his or her physician may interpret the cancer, in many ways. But if the cancer is hereditary and shown to be so through genetic testing, there will be the likelihood that the cancer will be at least as severe as that which the patient's progenitors had.

The Myths of Genetic Discrimination

Perhaps the fear that most animates those who have had genetic testing or who want to see it carefully controlled is a fear of what has come to be called genetic discrimination. One of the most important social changes resulting from twentieth-century genetic research has been activism to prevent discrimination on the basis of genes. A sizable majority of the world's nations have laws in place to protect the rights of those with diseases that are inherited. These laws are most often written to thwart the use of genetic tests in health and life insurance or employment evaluation. At first blush it seems like a good cause. How unfair to be the victim of discrimination merely because one has bad genes. How terrible to lose life or health insurance because a disease is "preexisting" in one's genes.

In the 1980s and 1990s, study after study failed to show what could reasonably be termed "genetic discrimination," while overwhelming evidence piled up in the United States that there is extraordinary discrimination against those who do not have health insurance, or who want health insurance but have a preexisting condition at the time of their application. Nonetheless, the fear that those who submitted to genetic tests would eventually be stigmatized set legislatures in motion. Many passed such laws as a kind of compromise in the war over preexisting conditions in general: allowing insurance companies to use information about preexisting conditions to set rates, unless those preexisting conditions were in some way turned up through a genetic test.

But these laws are based on two mistakes that characterize the twentieth-century confusion about genes: the assumption that *genetic* disease and *inherited* disease are identical, and the assertion that inherited diseases are *blameless*, while noninherited disease might involve *fault.*

Aiming at the protection of those who "merely inherit" their disease, lawmakers have discriminated against those whose diseases are not hereditary, or who cannot yet link a gene to their disease. In most of the United States, for example, a patient with prostate cancer who has not been tested for a gene, or whose test did not turn up a disease-related gene, could be denied insurance, while a patient with prostate cancer and a disease-linked gene will be protected by law from rate increases or insurance denial. It is silly to think that the patient with no gene is more at fault for his cancer than is the patient who inherits a gene.

Moreover, it is preposterous to claim that because no gene has yet been identified for a hereditary risk of cancer for the former patient, he in fact does not have a genetic disease. Why? In the twenty-first century it is clear that every bacterial infection, every virus, and every cancer work by making changes in the genes of human cells. Viruses are nothing but genetic information, most often put together such that they can harness host cells by reprogramming their genes. Cancer occurs for a variety of reasons and in a variety of ways, but it is a genetic event whether or not it is inherited. Thus the idea that a "genetic test" must be—or even usually is—a test for an inheritable condition is a misnomer. Cancer, and virtually every other disease, is genetic in the sense that it happens at the molecular level in ways that involve a change in the genes of the patient's cells. The problem, of course, is in the notion that genetic disease must be inherited, and through a specific mutation of a specific gene. This is wrong, and the idea that regional and state legislatures or judges will be able to sort out the difference between inherited genetic disease and that which is caught, like a cold, is just another fiction of twentieth-century genetics.

In geneware, one begins with the assumption that there are hundreds or thousands of inherited risks, all of which are in conversation with the environment and with the willful actions of the person toting around the genes. Mapping those risks is a matter to be entertained carefully, but one thing that map will assuredly mean is that insurance as we know it today will not be able to systematically exclude genetics, at a minimum because no one in an insurance company even *claims* to know how to deal with genetics in setting premiums.

In the 1960s, genetic testing was associated most strongly with the test for Down's syndrome in fetuses. Down's is caused by an extra copy of an entire chromosome. Clearly a test that finds an extra chromosome is a genetic test, even if it does not look for a specific gene. But any test that looks for changes in genes or even for the effects of a disease that works at the molecular level could reasonably be called a genetic test. An attempt to restrict genetic testing in insurance to tests for "hereditary susceptibility" would not work because even a flu test might turn up a hereditary susceptibility, given the pace of research in the genomics of the human immune system. Would a law that prevents genetic discrimination protect you against an HIV-test result? Should it?

There can be no question that eventually there will be discrimination of some kind, and probably of many kinds, on the basis of genes or even the genetic differences between populations that only show up as SNPs. One case in 2001 illustrates the possibility. Burlington Northern Santa Fe Railroad conducted genetic testing of 18 of the 125 employees who had filed claims in the year 2000 for work-related injuries based on carpal tunnel syndrome, and at least one employee alleges that he was threatened with firing if he did not provide the sample. Whether or not the chromosome 17 deletion tests that the railroad was using could ever be useful in predicting a hereditary risk of carpal tunnel syndrome, there is no question that the rights of an employee would be violated by a company's insistence on a medical test, especially if that test were used in promotion or exclusion from insurance. But when is discrimination *discriminatory* in the pejorative sense?

Blame, Blame

The debate over genetic discrimination is focused on the wrong problem. While the principles of individual liberty and equality are at stake in the same way they always have been, the development of digital technologies means that a tremendous amount of information that might previously have been considered discriminatory because it was inherited rather than "fault-bearing" will now fall into the same category as physical capacity and mental capacity and character and even judgment. Indeed, genetic information about all of those things will be in evidence within a matter of years.

Alarm about genetic discrimination will be muted when that discrimination would prevent a person who can be shown to have a condition that prevents him or her from making split-second decisions from driving a fighter plane, commercial airliner, or freight train. Burlington Northern Santa Fe might be the best test case for how the public would view genetic testing for the presence of viruses or inherited mutations that impair vision at high speeds, or that cause certain kinds of illnesses that render the employee unable to safely operate heavy machinery, or that make an employee an especially good candidate for psychotherapy.

As the range of choices about how to run a home, an office, a project, or a family expands, one new element will be those choices that involve assessment of genetic potential, or assessments of specific inherited risks. Few will doubt that these kinds of tests should be amalgamated into the testing that is already performed by lovers, by employers, by states issuing marriage licenses, by schools, or by major corporations and governments. Does anyone object to the genetic testing of future spies by the CIA to determine whether there is a known psychologically relevant risk?

The regulations about genetic discrimination will have to be expanded to prescribe remedies for those who are tested in this way, remedies that are intelligent and aimed at the expansion of liberty but that recognize that genetic testing is no unique threat to liberty, rather an extension of the range of information we already gather about one another in a free society.

The only real place for a ratcheting up of rules about genetic testing is back where we started our discussion—with Karen. Her genetic test was probably accurate, but *how* accurate does it have to be before a life insurer, employer, or physician can deny her something, or require and even encourage her to act? In order to determine the answer to this question there must be rigorous new standards for the effectiveness of genetic tests. Although a number of panels have been convened around the world to look at this issue, none have dared to suggest something as brazen as long-term trials of potential genetic tests before they are introduced to the clinical world. To do so would be to impede the impressive stampede to patent and market new genetics tests, and to ask national governments to fund a very ambitious new regulatory program.

For the past five years I have served on the U.S. government panel that is charged with evaluating genetic tests and devices, and in that period the panel has met three times, though more than four hundred new genetic tests are available. The Federal Food and Drug Administration, which has the authority to regulate genetic tests, does not do so because the tests are what is called "home brews," meaning they are produced in a single organization. Brilliant, huh?

Rules must be promulgated that will bring genetic tests—all of them—under governmental testing and oversight, so that no woman has breasts removed and no man undergoes chemotherapy on the basis of a false finding from a genetic test. Nor should any person forego a genetic test because she is under illusions about how ineffective a test may be, or about how simple it will be to fix a genetic disease later with gene therapy. But making these new rules means caring. If your congressperson isn't interested in genetics, fire them.

The Future Is Now

As genetic information becomes more available and at an increasingly early stage, genetic tests will no longer need to be part of the traditional medical community at all. One of our research groups has been examining the increasing use of e-mail by physicians, which has been hampered primarily by the physicians' fear that e-mail is too insecure or open to interpretation. As these fears fade, pushed aside by the demands already experienced by psychologists and social workers, some of whom now work exclusively by

Web and E-mail, there will be a great call to expand the potential for genetic testing on the Internet.

Part of this advance has been hampered by the need to develop genetic-testing hardware that will work in a consumer's home, but this is all but complete. Prominent computer genomics labs promise that DNA chips will be ready for use outside of the automated lab within three years. Your laptop or even your PDA may turn out to be the perfect way to test your genes: a disposable card takes a usable cell from your skin and conducts genetic tests, the output of which will be calculated in terms of risk and benefit, and regularly updated on the Internet through the computer.

The other part is more dramatic—the need to keep that information secret, to control its use, and to obtain good findings from it. Those who seek new genomic information at home will initially look a lot like those who want genetic tests in physicians' offices. They will fear something, a disease or a condition that has risks for them and for their current and future family. But as geneware evolves, they will use genomic information in the way they weigh themselves or assess their biceps.

Home genetic testing will give birth to the enhancement of human capacity through the design of lifestyles complementary to one's genetic information, and to the balancing of risks and benefits using not only genetic tests but technologies that act as a crutch for any number of genetic failings, or as an amplifier to special skills.

Genetic Tests That Work

You can use geneware now. It isn't all in beta testing, whether it is regulated or not. Not every genetic test is a dud, and not every effort to develop a test is rooted in senseless greed. Genetic testing is, beneath and above its extraordinarily visible failures, the future of medicine in a nutshell. Tens of thousands of genetic tests will come into clinical practice within a decade, and the integration of genetic testing with medical and other life decisions will be less menacing than existing tests would suggest. A genetic test will be part of every checkup—but perhaps more important, genetic testing will slowly begin to shape how patients and their families think about risks and lifestyle. Genetic testing will give parents an early heads up about allergies and susceptibilities in children. In fact, it is in the realm of the everyday that genetic testing will have its most profound impact. Genetic testing will tell you what foods are likely to provide the best possible match for your metabolism and what time of day your body is best able to digest them. Genetic testing may not tell you whether you are destined to be obese, but it quite clearly will offer insight into the kinds of toxins and foods that are more dangerous to you than to others who do not have your genetic makeup—the parts you inherited *and* the parts you have acquired.

beyond genetics

The Century of Risk Analysis

Testing that offers a glimpse of risks one cannot avoid, or a hint at what kind of activities may minimize any of a variety of risks, will fit fairly easily into what has already become a culture of risk analysis. This is a society that has moved from mandatory but ineffective seat belts— roughly the same in every car—to hyping minute differences between the crash-injury risks associated with driving dozens of very similar vehicles with their sophisticated injury-prevention devices. If you would buy a car because it is 10 percent less likely to cause a knee injury in a forty-mile-per-hour head-on collision, you are an ideal candidate for a genetic test to determine whether you have a 10 percent greater-than-average chance of developing lung cancer if you live in a house with new carpet.

Genetic tests that work will be bundled together, and the software that is used to incorporate their powers of prediction into personal lives will look a lot like other software used today to keep track of and make projections about personal money matters (planning for children's college or for retirement). Much of what is discussed today only after a person becomes a patient—with the symptoms of a disease—will be thought of in the way that one thinks of personal nutrition, hygiene, and fitness: as a matter of prevention and self-knowledge. Tracking your genetic inheritance will no longer be a matter of family trees. Software will let you plan based on risks and aptitudes that are encoded in genes.

In the *Handbook* of Epictetus, distributed to Roman soldiers thousands of years ago, it is advised that every prudent person take a long, hard look at himself before choosing a career. It doesn't make much sense to stake your life on Olympic running if you do not have that sort of musculature. The American mythology of overcoming physical challenges through raw, puritanical exertion will be with us even in the era of genes: The movie *Gattaca* tells the beautiful story of a man who overcomes "bad" genes to become an astronaut, a feat that should never have been possible given the genetic predictions at the time of his birth. But the beauty of taking risks not licensed by genes need not be eliminated by the possibility of genetic self-knowledge. If one sets aside for a moment the questions of genetic discrimination to which I have devoted so much time in this chapter, some pretty impressive possibilities emerge: greatly enhanced life and the power not only to know but not to know about hazards and advantages of many kinds. That knowledge would in all likelihood not only lead individuals to make choices that prevent danger but also lead societies to eliminate hazards that are shown to damage humans and animals at a genetic level. Genetic testing may even be the key to demonstrating the dangers of pollution and certain kinds of industrialization. But one thing genetic testing has not done so far is to enable the development of what people believe is its most likely product: effective gene therapy.

chapter five

bugs in the
geneware: gene "therapy"

the tragic and very public death of eighteen-
year-old Jesse Gelsinger in 1999 resulted in an all-out
halt of the most prominent kind of gene therapy. Jesse
wanted gene therapy because he couldn't keep his rare
disorder in check; the diet he had to follow was too
complicated and oppressive, and slowly he was beginning to suffer from a disease that had already killed his
younger brother. He wanted, as he put it, to save the
kids with his and his brother's disease.

But something went terribly wrong.

Jesse Gelsinger's death was one of hundreds that
would occur in the midst of clinical research in the
United States that year, including more than a dozen
in seemingly innocuous fields like dermatology research.
What made Jesse's case different was that it was the first
documented death of a subject in gene-therapy research

that could be clearly tied not to the underlying disease but to an adverse event caused in some way by the experiment.

Great hope and fear about gene therapy has been around as long as the term, but when it was personified by the death of a young man who was barely ill, everything in gene therapy suddenly seemed to just about everyone, including journalists who months earlier had been calling for dramatic expansion of gene-therapy research, to represent a kind of launching pad failure of a great space mission.

Before Jesse's death, many patients and most major American medical centers had invested a lot of hope and money in gene therapy; it has defined the ultimate hope of the genetic revolution for decades. But what is it exactly that gene therapy offers to Jesse and the thousands of others who enroll in gene-therapy trials? Most gene therapy is really experimental, designed to determine whether or not genes can be transferred into human research subjects using viruses without causing deadly reactions. Most gene-therapy experiments are misnomers, offering no real potential for therapy for decades to come.

Gene therapy that is more effective than conventional medical therapy has eluded science despite huge governmental and corporate investments, but this has not stopped the development of massive gene-therapy programs. Gene therapy is the goal most Americans believe will be achieved by the banking of genetic material and the development of genetic tests. But it is easy to confuse the promise of genetic tests with a future in which every genetic defect is fixed by lifting out bad genes through gene therapy. Just as genetic testing seems to promise a future in which genetics will allow us to customize our lives, one would guess that it is only a matter of time before genes can be rewritten so that no one need die of a hereditary gene mutation.

But gene therapy is not yet efficacious, and it can be deadly.

Jesse's Death

Shortly after I arrived at the University of Pennsylvania I was asked to participate in what my colleagues then described as our "fire station" role. Penn had the largest gene-therapy program in the world. Money for gene therapy from the U.S. government and commitment from many scientists, companies, and investors had helped Penn launch the best-funded bioethics program in the world. One payback for that was the idea that while Penn would rely on its regulators to make sure it was in compliance with existing laws or ethical standards, it would also hire people like me to think more generally about ethical issues in gene therapy.

The very first gene-therapy trial in which I was to have a role was for a rare metabolic disease called ornithine car-bamoyltransferase deficiency (OCTD), a single-gene defect for which the gene has been isolated. Quite rare, OCTD occurs only once in forty thousand live births in the United States. The disease is characterized by the inability to break down ammonia efficiently, which leads to toxic levels of ammonia in the blood.

Approximately 60 percent of males with OCTD have a severe form of the disease. They appear relatively normal at birth and for approximately one day. However, within a day or two, symptoms of hyperammonemia (intolerance, vomiting, lethargy, respiratory distress, seizures) become apparent and can lead to coma.

Approximately 50 percent of these newborns survive this crisis period using dialysis. Of those survivors, three quarters have mental retardation or other developmental disabilities, and only half reach five years of age. Virtually none reach adulthood. The other 40 percent of males with OCTD and 5 to 10 percent of OCTD carriers (females with a mutation in one of their two OCT genes) have its less severe form. Some may be completely asymptomatic, while others have a chronic condition requiring constant treatment, including a nitrogen-restricted diet and medication to counteract the ammonia. The diet means restricting most forms of proteins and fat, and the medication regimen is severe even compared with the cocktail of drugs used to treat HIV: dozens of pills at odd hours, with many side effects.

The OCTD gene-therapy trial was proposed for a number of reasons. First, it presented an opportunity to determine whether or not genetic information could be coaxed from viral vectors—the viruses that insert needed genetic material into the cells—into a large number of the patient's cells in the liver, enabling the cells to function properly. Second, this would be the first gene-therapy trial to administer a genetically engineered (or *recombinant*) virus to a human being through the bloodstream, thus exposing the entire circulatory system of the patient—and all of the tissues fed by that circulatory system—to the virus. Third, while early (so-called phase one) trials are designed primarily to assess the toxicity of the drug or device under consideration, in this case it was thought that at some point it might be possible to administer the viral vector to the infant males who had the severe form of the disease, potentially offering the only possibility for recovery from their crisis period. This early trial would thus move right to use of the device under consideration rather than just assessing its toxicity.

The proposal for this trial came at an unusual time. The U.S. government had essentially dismantled its regulatory body for genetic devices, the Recombinant DNA Advisory Commission, or RAC, and authority had passed to the U.S. Food and Drug Administration. The FDA's concerns were very different, and it had meager staffing and funding to fulfill its charge to oversee all aspects of this new and very complex form of research.

A few colleagues and I were asked to write a paper about the ethical issues in this trial of gene-therapy research, along with those researchers who would devise and in some cases help administer the therapy to patients.

Early on it became clear that there would be tension between the ethics side and the science side. In the draft we developed for the never-to-be-published paper, my colleagues and I asserted that the most important problem with gene-therapy research was a profound potential for misconception among researchers and subjects about what is being attempted, and, more important, a real question about whether to call the research "therapy" at all.

It was alleged that the Gelsinger family did not know enough; that the trial should have occurred only with much sicker children in the first place; that Jesse had become just slightly too sick to qualify for the research. It was alleged that the scientists at Penn failed to tell Jesse or his parents about problems that had occurred in trials of the gene therapy using monkeys. More troubling, *The Washington Post* devoted a significant cover story to the investigation of conflicts of interest on the part of the head of the gene-therapy program at Penn, conflicts that would have, *Post* writer Rick Weiss claimed, held out a strong incentive to cut corners and push ahead too quickly with the trial.

Because the case never went to court, the world will never know whether a jury (and the court system) would have upheld these claims against the scientists, Penn, and gene therapy as a whole. It is also unclear whether or not Jesse was in fact the first patient to die in a gene-therapy trial, since many earlier trials involved fairly sick patients whose deaths might or might not have been hastened by the trial of, for example, gene therapy for cystic fibrosis.

But already it is clear that there were grave misgivings about proceeding with this or any similar gene-therapy trial. I believe there was no way that the OCTD therapy, or any of the similar therapies in trial around the world, should have been tried on patients. The reason, in my view, is that they fail to pass a basic test of early research in that they could not get out of the early stages of research without dramatic improvements in the mechanisms used to deliver the genetic information (the vectors) and similarly dramatic improvements in the technology that would keep the patient's body from attacking those viral vectors before they could be of benefit. In essence this could be argued to be basic science masquerading as therapeutics.

Gene therapy needs several things in order to work. Genes have to be transmitted into the cells of the organism, and those cells have to successfully integrate that new information and produce whatever product that gene is designed to make (*express* the gene). But the best transfer rates ever seen in gene-therapy research have been quite low. Gene therapy also has to achieve a change in the patient's health that can be sustained. Because researchers use a virus for this research, the immune system comes into play, attacking the infected cells or just destroying the virus. After a short time, the body acquires immunity to the viral delivery system as it would to any such virus.

In the case of the OCTD trial, I argued that if you cannot beat the immune system, you have no business giving the virus to patients unless there is no risk involved. In any event, it makes no sense to call the attempt a "clinical trial" because, barring revolutionary changes in immune suppression technology, there is no hope that the mechanism being tested could ever be used in a real therapy. In essence, you have to call the research what it is—a risky experiment—and it makes no sense to do a risky experiment of this kind on sick patients or in the liver until the principle of gene transfer and immune suppression can be tested safely in healthy volunteers.

Needless to say, the problems we raised did not make many people happy on the gene-therapy science side of the department. In fact, our cautionary tale about starting gene-therapy trials prematurely were deleted from the paper, which eventually went to the bottom of a file cabinet, since half of its authors could not afford to see such an evaluation of their practice appear in print.

Not anxious to spend the first ten years of my career jousting with opportunistic young gene-therapy stars, I turned my attention to my real love: the study of how genetic technology affects society. But only a few years later the death of Jesse Gelsinger, one of the second group of volunteers in the gene-therapy trial, would become the reason for me, my colleagues at Penn, and many other researchers at other institutions to rethink gene therapy and what it can and cannot do.

Jesse's death brought down gene therapy; it resulted in the global shutdown of clinical trials in gene-transfer research, and the emergence of a new kind of lawyer, who successfully settled with Penn for an undisclosed amount of money to resolve the concerns of Jesse Gelsinger's father about his perception that the study never should have been conducted in the first place. Even those gene-therapy trials that offered real promise in their present form came to a crashing halt, and with them potential cures for everything from sickle-cell anemia to "bubble boy" disease to premature baldness. The successes of many different kinds of gene therapy would be set aside for a time not because of the riskiness of gene therapy but because gene therapy had achieved a kind of Frankenstein-type horror among the townspeople, because it had become clear that it was being rushed to the clinic, and because journalists had cast the entire enterprise as profiteering and risky.

It has become a recurring theme for gene therapy: hype, followed by a bit of success, followed by disasters laden with hubris. Jesse's death would call attention to the ways in which universities and small companies have rushed genetic innovations to market, and perhaps have cut corners to please those who fund the research or to achieve dominance in a field that was once thought to be the cutting edge of what I have been calling geneware.

Gene therapy cannot be regulated without an entirely new approach to genetics in government; those who regulate the regulators must recognize that gene therapy is neither the most dangerous form of research nor a run-of-the-mill activity to be left to the discretion of professors. What is needed is nothing short of a completely new approach to regulation of genetics at the regional and national level in the United States and abroad. In the United States the first step has to be educating Congress through the creation of a special agency dedicated to genetics, and providing enough funds to the regulatory agencies so that they can begin to offer better oversight of genetic research.

Is Gene Therapy Right for You?

Would you be able to tell the difference between gene therapy that presents a real potential for benefit to you and gene research that is more likely to harm than help? The answer isn't simple. There are only a few gene-therapy supershops, special facilities that have proven that they can do some form of gene therapy so well that a trip across country is merited. But for every promising gene therapy in clinical trials, there are a hundred difficult problems to be conquered before it can be demonstrated that the therapy will actually work. Deciding whether or not to use a gene therapy isn't so much about having good judgment as about learning how to distinguish between clinical trials that are likely to merely produce an effect and those that are designed to benefit you. In the short term, the real mission of gene therapy isn't therapy at all, it is research—and most of that research is aimed at finding out when a gene therapy is toxic, or when it produces an effect, not at curing anyone. If such an early clinical trial of a gene therapy is your last hope, you are probably not the best candidate for the trial.

Not every hyped genetic innovation will work, and often the line between truly effective and merely attractive is hard to find.

That is true of the likely successor to gene therapy: stem-cell research. Stem-cell therapy is the most controversial technology imaginable, involving (in most cases) the manufacture and destruction of very early life, life many would call human and some would call a person, for the purpose of transplanting its developing cells into the sick.

As a political debate, stem-cell research is the whole kitchen sink, an improbable combination of the abortion, cloning, fetal-tissue, transplantation, gene-therapy, animal-rights and enhancement-technology debates, raising worries about women in research, sex, the regulation of IVF clinics, the danger of changing the human germ line, and the war against aging. The hype is there too; it could restore youth or cause cancer, repair traumatic injuries or destroy tens of thousands of potential lives. Before it is developed, some of the most powerful politicians on earth will find themselves forced to modify deeply entrenched views (already, leading pro-life Republican Strom Thurmond has made clear his support for stem-cell research on diseases associated with dementia), and a few dozen scientists will become billionnaires through patents on bits and parts of embryos.

Stem-cell research is heady stuff at Johns Hopkins University, where research into the most basic elements of human development is being conducted, but it is also sold as a snake oil–style transplant in a seedy Russian clinic. More than 150 million Americans and perhaps another billion around the world may be treated with it before the decade comes to an end, yet almost no successful human clinical research has yet been performed using embryonic cells. It commands the attention of the major newspapers, news media, and scientific and business press every day, yet not a single textbook has been written about it for scientists, and most scientists disagree about the most basic aspects of the research, down to the question of what to call an embryo used to produce the cells.

Stem cells could come from a number of sources, some more controversial than others. Every adult has some stem cells in his body, and the most potent stem cells are abundant in early embryos. Perhaps the most promising research is also the most controversial: using a cloning technique to make an embryonic copy of a patient, then destroying that embryo at five days of age to harvest perfectly compatible and perfectly healthy cells for transplant into the patient. However stem-cell research is conducted, the catch is that to cure Parkinson's or Alzheimer's, scientists will have to cross another invisible line, between making babies through high technology and manufacturing embryos for genomic therapy.

There *will* be effective forms of gene modification, and within ten years some will be as common as aspirin. Yet again, sorting through the various options that carry the moniker "gene therapy" will be as much about detecting hype as about monitoring clinical treatments and science. The key is knowing how to do that; how to research and fix one's own geneware using tools from the information sciences.

The Future

Twenty-first-century gene therapy is crude even by twenti-eth-century standards; its attempts to change the genetic information in cells through a viral delivery system will one day seem very much like blindly shooting at flies with a machine gun. The future of gene therapy is not primarily in changing the genes of patients so that they can work again, but in identifying those genetic techniques that can produce drugs and devices that are best suited to improve the life of the person with a disease, whether or not that disease is inherited.

Gene therapy might turn out to be a whole variety of things. Part of it may be psychotherapy, a lot of which will be group therapy. Most of the developed world has out-dated ideas about heredity and its implications, and the adjustments that will be necessary for society are not so much adjustments to new technologies as adjustments to rapid change; humans need new and more highly evolved skills for coping, tolerance, caring, and loving. Genetics, and particularly the power to change genes through technology, puts a strain on all humanity where these skills are concerned, and it is not difficult to conclude that anyone who makes one of these new choices will need more than just a quick course in genetics to adjust to the outcome of the decision.

Part of gene therapy will not involve any kind of attempt to change the genes of the patient. Already diabetics undergo a kind of gene therapy; the insulin that many take is manufactured using genetic engineering of yeast bacteria so that they produce human cells.

The next wave in such gene therapy is the engineering of crops, animals, and other foodstuff so that it carries either DNA or proteins that would otherwise be hard to manufacture or difficult to administer. Breakfast cereal is already practically a plastic product; how much of a leap is it from chocolate bits with infused synthetic vitamins to a breakfast cereal that boosts protein, cures baldness, or improves sexual stamina? Technologically, it is a big leap. But morally? There hasn't been a natural food in the grocery store in more than a century. It's all a question of how to understand the differences between food and medicine and preventive supplements, differences that fade a little more every year.

chapter six

geneware in your kitchen

i can still remember the first time I knew that I was eating something artificial. My dad took me to the Smithsonian Air and Space Museum, and there, next to a new and gigantic theater showing space movies, was a concession stand with astronaut goodies. We bought tiny, warm, freeze-dried ice cream sandwiches. They were *like* ice cream sandwiches, but not. They had the texture of glue and Styrofoam but the flavor of sand and sugar. Somewhere in there was the ghost of an ice cream sandwich. Maybe.

I grew up in the 1970s as part of a generation bombarded with advertising about food *products*. But I didn't recognize bologna, chocolate breakfast cereal, chewing gum, or even Tang as artificial. As far as I was concerned, this synthetic treat might have made for an ideal way to get sweet stuff to the moon, but it didn't stand much of a chance in the ice cream aisle at the market.

The fight about how humanity should produce food has a long history that is both fascinating and disturbing. Food historians tell us that during the latter half of the twentieth century, the fight centered on three things:

1. the development of industrial methods for producing animals, fruits and vegetables, and food products
2. the proliferation of artificial preservatives, flavors, colors, and sweeteners
3. the modification of animals and plants through genetic engineering

It is safe to say that despite a sizable worldwide political movement that rose up to oppose the food industry and the changing composition of foods, that opposition was ultimately unable to prevent a total transformation in the grocery store. The vast majority of food purchased in most developed nations has been flavored and preserved and packaged to the point that it looks, smells, and even tastes radically different from food of even ten years ago. And food is perhaps best known, of course, by its name and branding. Even bananas are identified by their sticker.

The modification of the genetic makeup of food is the most controversial aspect of the contemporary food fight. Genetically modified food has been a big issue for only two

decades, yet the breeding of new kinds of plants and animals has been at the heart of agriculture for as long as
there has been agriculture. Why so much worry now? It is
one thing, argue opponents of genetically modified foods—
GMOs (genetically modified organisms, for short)—to
breed a hybrid rosebush the slow way, and entirely another
to think of roses by another name: software.

The Album, the CD, Napster, and the Potato

A good way to understand the heart of this objection is to
ponder the difference between the vinyl record and the
CD. The analog record is, from the moment it is recorded
in a studio until it comes out of your home speakers, an
aural waveform. In the early 1980s Peter Gabriel stood
before a microphone whose transducer would vibrate in
just the same way as an eardrum, receive waves of sound of
differing amplitudes or frequencies, and amplify those
waves to a still larger mixer and amplifier. The result was
electrically recorded as waves on magnetic tape, then as
waves on a steel press (much like a printing press) that was
pressed over and over again into hot vinyl, impressing on
that vinyl the form of waves of sound still in the same basic
format in which they were recorded. At the end of the
chain is the consumer, whose record player has a needle
that bounces its way along the grooves of vinyl that contain
the waves, passing that signal to an amplifier so that you

will hear the waves as they come out of speakers that vibrate in just the same way as Peter Gabriel's microphone.

As a child, I used to put my fingernail on my dad's records as they would spin around on the record player. I could hear and even feel the music, quiet but "real."

A CD is software. But is it less musical? Is it worse? Is it artificial? In 2001 the same Peter Gabriel, among the first popular Western artists to embrace digital music, sings into a microphone. The microphone vibrates in resonance with his voice. But the electrical impulses generated by that vibration are immediately fed into a very powerful computer, which converts every bit of vibration into data: a 1 for sound in some particular frequency band, a 0 for the absence of sound. The computer has to create and record a lot of information very quickly, and it needs to be able to very quickly check the fidelity of its reproduction against the sound coming in. The formats for such recording vary and are changing constantly, at a rate roughly consonant with advances in the size and speed of computer storage and processing equipment more generally.

Compare a laptop computer from 1990 to a contemporary laptop and you get the idea: a digital recording from 1985 had perhaps 1/1000th the sonic information contained in a CD of today. When the CD is played in your home or car player, a computer called a digital-to-analog converter reads the data from the CD. It is difficult for the computer to process all of that information, so it too is "sampled," as the computer reads each few milliseconds of musical data several times, finally playing the average of the samplings. To play the data as music, 1s and 0s are converted back into vibrations to be amplified and sent out through the speakers.

But whether it is 1985 or 2003, the phenomenon of digital music (or for that matter of pictures and even smells) is fundamentally different from that of analog. If you understand these differences, you are well on your way to appreciating the choices that face you with digital food, the food that has come to be called "genetically modified."

Part of the difference is economic. There is no question which musical medium wins. Digital editing and production is infinitely more cost-effective. A digital sound can be added to a recording almost without cost. The digital recording can be reproduced as fast as your computer can process it. You can make your digital music files accessible to millions of people through the Internet, and the cost to each person who downloads a copy of your files is minimal. Every garage band can give its music to the world, and the cost of scaling up is strictly a production-and-marketing cost: making more copies of songs already produced is essentially free.

Analog music, by contrast, costs a fortune to print and package, and the only way to get a copy of an analog performance is to physically transport either the artist or another copy of a recording.

But the digital song may not be as faithful a reproduction as that made by an all-analog process. If you are in the
room with a pianist, in a good spot to hear the music, you
hear what the pianist hears and even feel some of what
goes on in making the music. A great analog recording
played on a great record player with a great needle and
great speakers has much of that quality; there is no "sampling" loss, because the vibrations of the music are all there
on the vinyl (unless you scratch it or your needle bounces
too little or too much). Such a process is incredibly expensive from start to finish, and each consumer who wants to
have the experience must pay a significant setup cost (good
tickets at a good venue, or a twenty-thousand-dollar stereo
system). A great digital recording may be almost as faithful,
but consider that by the time the recording reaches your
ears, it has been converted several times from analog to digital and back to analog, losing some of the original sound
each time.

But does a CD played back on an average stereo with
average sampling capacity sound better than an average
vinyl record played back on an average stereo? As *Billboard
Magazine* would put the matter in 1998: yes, with a bullet.
The album is dead for all but wealthy audiophiles. The
world of music has become the digitized orange: consumers who cannot afford to have an appetite for perfect
musical reproduction have come instead to love all the
tricks you can perform on music through digitization. Most

of those who long for perfect recordings are listening to four-hundred-year-old compositions; the miracles of modifying music as software are not part of their idiom anyway.

Digital Food

Naturally, the debate about whether food can be digitized, reprogrammed, and reproduced is more heated than the analogous debate about CDs. The objection to digital music is usually about whether or not digital apparatus helps or hurts the attempt to produce a true reproduction of a live performance.

Everyone involved agrees that a good reproduction is better than a bad one. The debate centers on disputes about whether digitizing music or for that matter anything results in a loss of something, and if so whether that loss is just esoteric (experienced only by rich people with fancy stereo systems) or outweighs the incredible advantages associated with making musical media much less expensive to produce and transport, at good quality, to many more people.

In other words, both sides of the debate want to hear music that sounds good, but one side asserts that music can only be real or natural when it takes a particular form. It is not an argument that began with the digital-music file; hundreds of years ago religious groups split over whether or not God wants musical instruments played in houses of worship, and if so which ones are natural and which technological. The disputants who argued that food must always be tended by hand, that crops must not become products of industry, and who often argue today that food sources ought not be genetically modified share no such values with their opponents in the labs that make genomic food.

For the food purists, the aim is often to keep the production of food from becoming a way in which people are alienated from nature or God. Sure, there are anti-GMO protesters who lament the loss of "real food" or, like audiophiles with turntables, the taste of a food that has somehow avoided human modification. But more often, the longing is less for a bucolic food of yesteryear than for a different way of life. Meanwhile, those who support GMOs aren't trying to get a digital duplication of the perfect tomato, or to bring back an ideal if Jurassic Park–like potato through the mixing of DNA. They have other economic, aesthetic, and scientific goals: the improvement of food through changing the basic features of plants and animals.

This fact could not be more hidden. Newspapers regularly report that those who oppose GMOs are marching against the replacement of natural food with high technology. It is a fear, we are told by writers like Bill McKibben, of what he calls "the end of nature." Leon Kass writes of the importance of finding the essential natures of food so that humans can live a "more upright," more truly human life.

By the same token, the arguments in favor of GMOs

are almost always made in the news media by those who are affiliated with labs and companies who are trying to make cheaper food through genomics. Critics say the scientists' arguments for GMOs are suspect because they originate in a quest for profit. But poll after poll has shown that people outside the U.S. and western Europe would welcome *safe* improvements in the taste of food through genetic engineering, and that some of the most desirable improvements are among the highest-tech: grains engineered to contain lower cholesterol, cattle engineered to lower the presence of heavy metals in the meat. Indeed, some changes in the "input traits" (those features of genetically engineered foods that affect only the efficiency or cost of production, as opposed to "output traits" that affect flavor and other consumer-centered matters) are not bothersome to those who see some delicious foods as entirely too expensive to produce and consume.

If those who oppose and those who support the idea of genetically modifying crops have little in common, and if the debate about the artificial is at the core of the debate yet is misleading in many ways, then what is the food fight really all about? The answer is predictably deep and fascinating.

The Rationale for the Fight

Food is central to culture and religion. The sharing of food is often viewed as an intimate event in which important promises are made, loyalties affirmed, and special relationships established. Most religions, researchers Dudley Burton and Daniel McGee note, mark the important events of their lives by feasting or fasting. Foods take on powers in many rituals to enhance or deprive the human body of sexual, physical, mental, and spiritual powers.

Burton and McGee also note that both sides in the GMO debate claim to be defending absolute and unchallenged values about these important matters; thus they tend to be absolutists in their claims and in their tactics. The food fight is a kind of holy war in which only complete defeat of the other side is acceptable, and in which every infraction by the other side is viewed with intolerance.

If food were simply a matter of cultivating one's own backyard, the debate about how to engineer food sources would be much less divisive. But food is international, and the many peoples of the world represent widely divergent views about how food should be grown, traded, and exported or imported.

What Have We Grown?

Currently the most important GMO crops in terms of acreage planted are soybeans, corn, cotton, and canola. In 1999 the area planted to GMO varieties was almost half of the U.S. soybean acreage and about 25 percent of the U.S. corn acreage.

Worldwide production area of GMOs and traits

Crop	Area planted in 1999 (MILLIONS OF ACRES)
Soybean	53.4
Corn	27.4
Cotton	9.1
Canola	8.4
Potato	<0.3
Squash	<0.3
Papaya	<0.3
Trait	
Herbicide tolerance	69.4
Bt insect resistance	22.0
Bt + herbicide tolerance	7.2
Virus resistance	<0.3

Source: McGee and Burton, 2003.

Globally, the United States is by far the largest producer of transgenic crops. In 2000 it had 74.8 million acres; Argentina had 24.7 million acres; Canada 7.4 million acres; China 1.2 million acres; South Africa and Australia less than .5 million acres each.

The two critical rationales for GMO crops to date have been to reduce the cost and difficulty of producing crops rather than the improvement of the crops. The emphasis has been on herbicide tolerance and insect resistance.

Genomic Weed Whacking

McGee and Burton note, "Weed control is one of the farmer's greatest challenges in crop production because poorly controlled weeds influence crop yield and sometimes quality as well." Herbicides of today are varied and sophisticated, but each can control only certain kinds of weeds and sometimes only at particular stages of a plant's growth. Herbicides also leave behind a residue that may cause human and animal reactions in the short and long term.

One use for genomics in the production of food is in making a crop that is herbicide-tolerant. This means that a strong herbicide can be used, one that would kill the wild (nongenetically modified) type of the crop, to greater success in eliminating weeds. Many of the broad-spectrum herbicides also break down quickly in the soil, rather than remaining as a residue on crops or in the water that flows from irrigation into the water table to be consumed by people and animals.

Genomic Bug Wars

The best known use of genetics to kill bugs is the incorporation of a soil bacterium called *Bt* or *Bacillus thuringiensis,* whose spores contain a crystalline protein. Bugs that consume *Bt* convert that protein into a toxin that dissolves the walls of its intestines. Genetic engineering of plants to splice in the genes of one form of the *Bt* bacterium causes exactly the same effect when pests consume the engineered plant. If the crop kills its own pests, less chemical pesticide is needed. And chemical pesticides can be highly toxic and unstable, not to mention extremely expensive to apply. But what does *Bt* do to humans? To date no study has demonstrated that it is hazardous to humans. But when corn engineered to contain one of the proteins (Cry9C) that had not been approved for human consumption, *Bt* was accidentally mixed with food corn for humans and wound up in food products such as Taco Bell taco shells, reports of allergic reaction spurred activists on both sides to make extraordinary claims about the healthfulness or toxicity of *Bt* corn, and its GMO cousins, to the world.

Better Food Through Molecules

A major concern when the future of GMOs is discussed is *flavor*. The best test case for flavor has been the genetic modification of tomato plants. The problem is that the super-tomatoes are not very successful when it comes to sales: the flavor changes just have not been persuasive to consumers, especially when they are presented as products of genetic modification. But where the tomato has failed, countless other experiments are in progress involving fruits, vegetables, animals, and mixtures of these with one another.

If your food can't be made more tasty with genomics, it

can certainly be made more healthful. Food and drugs have never been truly distinct, and the effort to combine them through so-called nutriceuticals has always had genomics at its core. Edible vaccines could save lives where the administration of traditional vaccines is prohibitively expensive. Transgenic bananas have been engineered to carry deactivated viruses that cause cholera, hepatitis B, and diarrhea, effectively serving as a vaccine for all of these diseases.

In animal agriculture the primary focus of work has been on using transgenic animals to produce desired bio-chemicals that can be isolated as medicines. The use of hormones, rBST, for example, to stimulate enhanced milk production in cattle, has been highly controversial. In addition, research is going on with "supersalmon," com-bining the genes of this highly prized fish with those of catfish or other lower-quality but faster-growing fish to achieve the desirable features of both species. These and other transgenic animals are rarely available to consumers, but it is just a matter of time, regulatory catch-up, and public perception.

Eating the Future

There are a number of issues to consider before you set the table or raise the flag of concern. Most are neither obvious nor simple. But none are avoidable.

1. Direct Health Hazards

Some individuals are allergic to certain foods, and these reactions can range from mild to severe. There are some examples of individuals becoming sick as a result of consuming food containing genetic materials from sources to which they are allergic. These concerns have led to calls for extensive precommercial testing of products and for more extensive labeling of genetically modified components so consumers can make more informed choices.

It appears that some GMOs may pose human health risks when consumed. For example, a project to insert a brazil-nut-protein gene into a soybean was halted when early tests showed that people allergic to nuts reacted to the modified soy products. This example demonstrates the need to consider potential health effects in new GMO crops, but it also shows that proper testing can identify risks. Although almost half the U.S. soybean crop and a quarter of U.S. corn now consists of GMO varieties (which means that we have all been eating transgenic food products for some time), there is as yet no undisputed case reported of anyone suffering health effects as a result.

2. Risks to the Environment

Bt corn, which contains a bacterial gene enabling the plant to manufacture a substance toxic to the larvae of pest insects, also kills the larvae of nonpest butterflies and moths. *Bt* has been a target of criticism since a laboratory study showed that *Bt* corn pollen dusted onto milkweed leaves was harmful to monarch butterfly larvae feeding on them. Follow-up studies have shown that pollen from *Bt* corn rarely reaches toxic levels on milkweed in the field even when monarch butterfly larvae are feeding on plants adjacent to a cornfield. Planting *Bt* corn also greatly reduces the need for spraying with pesticides, which are far more damaging to the "good" insect populations.

3. GMOs Blowing in the Wind

More information is needed about the extent of the risk of "genetic pollution" for different crops. However, a recently completed study at the University of Maine found that cross-pollination of conventional corn by transgenic corn grown in an adjacent plot was 1 percent when the plots were one hundred feet apart and declined to zero at a distance of one thousand feet. This result suggests that it will be quite feasible to prevent the transfer of transgenes to nontransgenic varieties by following recommended planting distances, just as is currently done to maintain purity with conventional varieties or to protect manipulated hybrids. Nonetheless, the genes do move, whether across short distances or long, and they find their way into neighboring crops in ways that may not have been measured yet, such as being toted about by us humans.

Where GMO crops containing an herbicide-resistant gene grow alongside closely related weed species, there is reason for concern about gene mobility. Proposals to reduce the risk of creating transgenic "superweeds" include linking herbicide-resistant genes to other genes that are harmless to the crop but damaging to a weed, such as genes that affect seed dormancy or prevent flowering in the next generation. Thus, if a weed did acquire an herbicide-resistant gene from a transgenic crop, its offspring would not survive to spread the herbicide resistance through the weed population. But by the time superweeds are identified, will it be too late to slow their growth through such measures?

4. Corporate Takeover of Food Production

In addition to the biological and environmental threats discussed above, there is concern about the potential social and economic consequences of genetically modified food technology. For example, one of the world's largest agricultural biotechnology corporations was awarded a broad patent for the use of antibiotic-resistant genes. With this patent, that company stands to benefit financially from, if not control, key activities in the whole world of biotechnology. As we will discuss, there are implications for this kind of patent that run across all of food and medicine. There is considerable anxiety among some observers about the pattern of powerful corporations controlling other processes of biotech development, anxiety that may be at the heart of the European objection, for example, to GMOs. For example, in the United States, the vast preponderance of patents for GMOs are held by private companies. McGee and Burton note that among the top patent holders, all but the U.S. Department of Agriculture, the University of California, Cornell University, and Iowa State University are private corporations.

Top Ten U.S. Patent Holders in Agbiotech, 2000

Monsanto	287
DuPont	279
Sygenta	173
Dow Chemical	157
U.S. Department of Agriculture	102
Aventis	77
University of California	48
Savia	33
Cornell University	33
Iowa State University	29

Critics sometimes play on public ignorance and fears about biotechnology to make ideological points about corporate control and influence. But what is the proper role of big business in designer breakfast? It is a question whose answer must in part depend on how one feels about the relationship between government, industry, and society, which we will discuss in the next chapter.

Whoever owns it, genetically modified food will be

judged by how it improves social life. Foods that are altered by genomics only to make them cheaper to produce are unlikely to warm the hearts of consumers. Foods that are the product of imaginative ideas about how things ought to taste in our mouths—genetic mixes of chicken and chocolate—are likely to be consumed only by those who like the idea.

Unless you are persuaded that GMOs are the end of the earth, it makes a whole lot more sense to evaluate the GMO issue one food at a time while learning as much as you can about the history and future of food for your family and society.

chapter seven

geneware is not
free anymore

before my friend karen could have a genetic test
for breast cancer, Myriad Genetics set the terms and
the price. Karen's doctor's lab in Philadelphia couldn't
do the test. In fact it had been specifically threatened
by Myriad with a lawsuit should it attempt to perform
any BRCA genetic testing. The BRCA breast-cancer
gene, discovered by a scientist working for the Human
Genome Project on U.S. federal grant funding, was
patented when the scientist made a mad dash to the
patent office before he published his results. Today he
has left his cloistered role as a scholar in a major university to assume leadership of the company that best
epitomizes the world of genetic intellectual property;
more and more of the genome is exclusively licensed
by companies and unavailable for research without
fees. So Karen's doctor will mail a credit card number

and a swab of cells from Karen's cheek to Utah, where Myriad will conduct the test and send back the results along with some materials that explain them. The test will cost her upward of three thousand dollars, a price that reflects nothing but the fact that Myriad can charge whatever it wants for as long as its patent lasts.

Seem a little strange? Welcome to the future of genetics.

DNA, whether it is found in your body or that of a cat or wheat field, is being gobbled up like the prairies and mountains of the great American West by prospectors intent on staking a claim that will eventually make them rich.

There are boomtowns in this great rush, every bit as odd and romantic as those of the gold rush. Reykjavík, Iceland, is where one company, sanctioned by the state, sifts pans of the genes of most Icelanders. Drive through Silicon Valley today and it is immediately clear that silicon is not the primary focus of computer entrepreneurs; the speculation is all about genomics and the tools and results of high-speed analysis of genes. Just as in the boomtowns of old, the only thing holding together much of the massive effort to turn genes into gold is the faith of thousands of prospectors and the money of some pretty aggressive venture capitalists, creating companies that trade at pennies or just a few dollars a share.

Big gene-mining companies will replace the entrepreneurial effort eventually, but in the meantime the ground rules are being set for people like Karen; the first one to map the gene gets to set the price and make the rules about its use.

Why can't Karen just go to a college genetics lab and do a BRCA test? The equipment is available, and most of it is inexpensive. The answer is the patent; you do not have the right to look at many of your own genes. You can borrow the right to analyze them or, more correctly, *license* the use of a patent to analyze them under specific and very limited circumstances, from the companies and universities that have placed patents on their discoveries of human genes. But the patent holder sets the price, which is most often quite high. And forget about challenging the right of a company to own the genes that occur quite naturally in your body; in the more than a decade since human gene patents became a fixture in European and American patent offices, no court has held that such patents are unconstitutional.

When the early-twentieth-century U.S. Congress took up the question of industrial devices, it ruled that anything under the sun made by man could be "patented." The holder of the patent was entitled to exclusive control over the device for a limited period, providing that he demonstrate the utility, novelty, and workings of it. Virtually every other nation has similar provisions for the ownership of novel devices. Courts in most nations have held that biological devices are patentable as well, subject to the same standards. Of course, applying the standard of novelty to the genes of an organism, or even to a device for ascertaining in an exclusive way any of the effects of those genes, is a complex matter, one that has not been carefully tested in the courts of the world.

What does it mean to patent parts of the human genome? Patents are designed to protect inventions: "anything under the sun *made by man.*" When genomics companies identify genes, what have they made? Is finding a gene more like finding a plot of land or underground water source, or is it merely inventing a way of testing for disease? The public has challenged gene patents in poll after poll, yet patent offices around the world have granted tens of thousands of patents on plant, animal, and human genes.

If gene patents make good business sense, how much responsibility do the owners of my genes have to tell me about what parts of my genome are off-limits? How am I to find out who owns my genes—who owns the risk of cancer I inherited from my mother?

If someone owns my genes, what does that mean for my identity? For my religious beliefs? For my right to privacy? Many of us today are hesitant to give out our credit card numbers online; how will we feel knowing our genes can be bought and sold, transmitted, used, exploited without our consent? In the past seven years many patient groups have urged children and adults with rare genetic diseases to give gene samples to researchers at major university labs so that genetic tests can be found. But in the past three years many university-based genetic-testing research programs have in effect become clients of the corporations who own tests that the university labs once conducted for free. Moreover, the companies who are replacing research-based university efforts, efforts that lead in most cases to the patents on genes, often return no profit from the genetic tests they find to the people who gave them the samples that made it possible. No scientist or policymaker expected that genes would be patented at the current pace, and if these trends continue, the entire human genome will be grabbed up as quickly as oil leases in East Texas a hundred years ago.

Who Cares?

Patents are not regularly the topic of breakfast-table conversation. Sure, your breakfast cereal probably contains patented vitamin supplements, and there is no question that your vitamin D milk is produced under a patent that earns the University of Wisconsin a mint. But why should you care? Patents protect inventions, and they may inflate the price of an inventor's new idea for a time. Those fancy stick-on blackhead removers don't cost twenty dollars a box because the materials are expensive; think of fifteen of that twenty dollars as a patent tax levied by the inventor while she still controls the innovation (in the United States, patents last twenty years from the date of filing). The inventor's reward has constitutional roots, and its importance is as well established in law as in basic business practice.

But at your table, not one of the patents on the products within reach is noticeable to you as you eat your breakfast. At the beginning of their life span, products based on patented innovation are more noticeable because (as is required by the patent laws in most nations) they are both novel and useful. You will remember the first time you eat a banana in the shape of an orange, and not only because it will taste and look unusual—it will also cost a lot more than typical bananas or oranges, and there will be a label on it with a patent number. New patents are most often useful for little more than lining garbage pails. But the few that really revolutionize things are noticeable not just because they change things but because it is just plain fascinating that an idea can be so original that it merits patent protection: "Someone came up with an orange-banana, and gosh, I'll bet they're rich!" The rest of the time, the fact that many of the devices you use are under patent just isn't all that interesting. I pass the juice without comment, even if it costs a few cents more because of a ten-year-old patent on the container. But if I tell you over breakfast that you do not own the right to look at your genes—and that someone else does—you'll pay attention.

A New Kind of Science

The distinctions between twentieth- and twenty-first-century genetics really depend on an understanding of the "operating system" of genetics, the framework within which scientists, policymakers, clinicians, and the rest of us understand what a gene is and what genomics is for. What will play the greatest role in shifting our mind-set about genetics is the emergence of new tools to put genetic information to even more revolutionary use, and a new vocabulary that will emerge from the use of those tools.

The danger of commercialization in human genetics is often attributed to the new practices of securing venture capital and pursuing intellectual-property rights, practices that come with market forces in biotechnology. The agricultural-genetics companies, most of whom have now been swallowed by larger chemical or food companies, led the way after the critical decision by the U.S. Supreme Court decades ago to allow the patenting of engineered life-forms. This gave companies something to own and something to make and something to sell, and converted science into technology in the commercial sense of that transition. No doubt, commercialization is changing genetics. In some ways this change has been amazing, yielding a massive increase in published work by those who work in technology transfer labs that team the power of the university with the money of the corporate world. Much of the development of the American and Western university in the last twenty years has been financed by biotechnology and pharmacology profits, and in various ways the products of biotechnology and genetics have improved the health of millions of people and in many ways eased the lives of the poor.

The objectionable changes, though, are truly foreboding. They are linked to what people value about science in the first place. Perhaps no theoretician of the ideals of science is as widely respected as Thomas Merton, who described several that merit mention:

1. disinterestedness—the willingness to share the results of research without promise of economic reward
2. communalism—a commitment to the community of science as manifested by openness of communication and aversion to secrecy
3. freedom—to pursue interesting but unprofitable research, or research whose profit yields are not yet known
4. review—the commitment that publication should happen in peer-review journals

Recent studies have shown that whether or not some scientists practice Merton's ideals can depend on who pays their bills. In one national study of scientists conducted by Harvard University in the 1980s, 32 percent said that the likelihood of commercialization influenced their research topics. A 1998 study of scientists in major British research institutions put the figure at 53 percent. And then there are the well-known instances in which pecuniary drives overtook idealism entirely, such as in the 1989 case at the University of Pittsburgh in which a professor altered his interpretation of results of clinical trials after fees were paid to him by his funding organization. Many in science agree that industry has changed science a lot and exercises much more control today than even five years ago. It can restrict the publication of good but boring or controversial science, it can result in articles authored by industry scientists but attributed to scholars who desperately need publication credits; it can pressure scientists to alert potential investors through the lay media prior to publishing peer-reviewed science.

Bad science in the lab can cost millions and put the integrity of science at large on trial. Bad science at the bedside can put patients in harm's way. Conflicts of interest at the bedside are unavoidable, a physician's salary will depend in some way on incentives to do more or do less, and countless others who participate in care of patients have multiple allegiances, which are all too often hidden from patients. Did the choice of an inhaler for Patient X stem in part from the fact that the company that makes the inhaler has been bringing a really nice free lunch to the hospital staff twice a month for the past year? In high-tech medical institutions, physicians are often researchers as well, and this presents a conflict between the role of physician as physician (a duty to provide the best care) and physician as researcher (a duty to conduct studies in an unbiased and accurate manner). To this, commercial interests in research add an additional conflict of interest as the physician becomes an investor in or employee of a company who sponsors his or her work. In the mid-1990s a prominent Boston Eye Clinic researcher made an illicit, midstream alteration in the design of a study involving his own patients to minimize negative findings *before* selling his 530,000 shares. Many gene-therapy programs have been accused of moving too quickly to the clinical phase of their work because of the pressure to satisfy venture capitalists' need for a product that can be marketed to investors, if not patients. And when Karen shows up at the doctor's office to inquire about a breast-cancer genetic test, she will not know about the incentives or disincentives that her doctor has been given regarding that test. For that matter, even the ethicists who assess genetic practice are often funded by corporate or government stakeholders, such as my funding from Iceland. Conflict is ubiquitous, and patents can turn small conflicts of interest into giant secrets that reshape the whole face of the future of health care. Genetic patents benefit from geneware: you can't hide a violation, because the computers and "data" are visible everywhere.

Gene Patents

In October 1999, Celera Genomics filed a preapplication for more than six thousand patents on genes it identified in its efforts to date, creating a debate in Congress, the media, and the scientific community about whether it is ethical to patent human genetic information, and whether it makes good commercial sense. That debate seems to be building steam as more of the world community sees the stakes.

How much should a particular investigator, be she professor or corporate employee, own? It is the Pete Rose question: should professors be betting on their own research, and if so, can we reasonably expect them to bet in ethical ways—or is it more likely that the timing or outcomes of their research will be shaped by their own stake in the research? Is it a conflict of interest to own one's research (as odd as that may sound), and if so, what is the root of the conflict? Whatever one may say about the era in which science and technology were considered wholly separate entities, that era has clearly come to an end.

Corporate money and influence are not always anathema to the successful transition of basic science into clinical remedies. Without massive investments by pharmaceutical companies in research, manufacturing technology, marketing, and other infrastructure, no chemical compound created at a university would ever make it to the drugstore. The university as an institution is not equipped (with few exceptions, such as MIT's incredible investment in a "technology park") to bring its own discoveries to market. Most Western governments are in even worse shape where selling things is concerned. Could a U.S. Congress that cannot agree on how to provide health care for millions of its citizens really be an effective source of funding for some kind of "federal" pharmaceutical-manufacturing effort? Not likely. Shortly before the twenty-first century, a whole range of legislation opened the door to investment by companies in research that is funded by government agencies. Universities, and in particular medical schools that work on genetics, have grown or faltered in almost direct relation to how well they take advantage of the new opportunities for public-private hybrid research. There is no question that a new era has dawned; the genetic test for which Karen will pay three thousand dollars is owned by a company whose research was conducted under a federal grant that Karen helped fund. Karen isn't looking for a taxpayer's discount. She just wants a square deal. If it is important to pay for science that is eventually owned by companies, that's one thing. If she is paying for an unnecessary or even unhealthy monopoly, it's another. Karen doesn't understand why a company should need to own her genes, with or without taxpayer aid.

Gene patents have ushered in a whole new understanding of what it means to discover, make, or possess things as a scientist. There is a big question about how to understand ownership of genetic research, and an even bigger question about the implications old-style policies could hold for society if they are not updated.

The Case of Patenting: Conceptual Problems Run Amok

There are two arguments against gene patents.

The first is that genetic information is a part of nature, and one ought not, indeed cannot, patent nature. Patents allow long-term control of new and innovative processes to be secured by inventors.[1] You have to figure out how to do something in order to receive a patent, and the process you create must have utility. Critics charge that finding a gene is not innovative. Arthur Caplan and Jon Merz, for example, compare the identification of disease genes to the land rush.[2] The basic tools utilized in the multimillion-dollar laboratories that identify genes, they point out, are themselves covered by patent protection, and render the search for genes similar to using conventional means to find new territory. Caplan and Merz contend that while the scientists who identified some forms of a gene that correlate with breast cancer are entitled to claim their discovery, discovering new land is not the same thing as owning a process. Put another way, while the telescope is indeed an innovative instrument, and someone can indeed patent the technologies involved in telescopes, it is another thing entirely to point your innovative telescope up at the heavens and begin claiming each new star as a product of technological innovation, thus protecting the process of looking at each new star.[3] The point is well taken. The person who discovers Pluto has not invented a process, no matter how much utility might be claimed from finding and using the discovery of Pluto. If we follow this line of analysis with genes, locating new genes is a matter of sailing into new territory with old boats, so the correct mode of protection for the "finders" of new genes would be something other than patents, akin to land-use or water-rights laws.

However, it seems to me that those who advocate this first argument against patenting are barking up the wrong tree. They assume that the U.S. Patent and Trademark Office, and the courts that enforce patent claims, can be a final arbiter for some arcane, metaphysical distinction between "discoveries of nature" and useful technologies made by humans. Opponents of patents, especially those patents for "methods of detecting" the relationship between genes and disease, assert that gene patents are improper because they involve no "reworking" of a gene from its "natural form," and are thus attempts to patent nature. They argue that genes are natural phenomena discovered using previously patented devices. Merz, Cho, Robertson, and Leonard, for example, write that "the discovery that a particular DNA sequence at a specific locus or that different forms of a gene are associated with a disease does not qualify that scientific knowledge as patentable subject matter because no human alteration of an existing organism or naturally-occurring entity is involved . . . [and] because it is merely an observation of a state of nature or 'nature's handiwork.' "4

But is it so clear? It seems to me that while disease genes are in one sense discoverable by conventional means, their utility and indeed their meaning as commercial objects is not discovered but rather invented. Investigators who patent a specific form—or even *all* forms of a gene for susceptibility to cancer can legitimately claim a patent for "methods for detecting," a relationship between a particular bit of DNA and some phenotype. The job of the U.S. Patent and Trademark Office is to evaluate whether or not the patent claims a novel method for detecting a particular relationship between the DNA of a person and some phenotype, for a particular purpose (e.g., diagnosing the presence of possibility of future disease states). If that utility cannot be demonstrated, a patent is not granted. Methods patents do not patent disease genes per se, but instead the process of making use of that DNA in diagnosis.

Genetic Innovation

There is a subtle distinction to be made between "observing DNA" and constructing a DNA-based product for diagnosis of some disease or phenotype, much too subtle to be captured in any obvious way by the blunt rules about patentable subject matter, let alone by the intent of our Constitution's founders. Nonetheless, the distinction is real, and goes right to the heart of what patents are supposed to do. Disease-gene patents are more an innovation of scientists than a discovery. That we sometimes tend to believe otherwise is to some extent a product of genetic essentialism, a cultural belief[5] that genes are a simple, self-evident library of data present in everyone and responsible for all aspects of human embodiment and disease. If we begin with the idea that genes are a simple code to be read or stumbled upon, we miss the immensely difficult epidemiological task of purifying otherwise diffuse relationships between particular environments and genes, and between particular groups and genes.

If I find a strange tree in the Vermont forest, take a clipping back to my lab, discover that eating the bark cures a disease, and file a patent application for the tree itself, or for very broad uses of the tree, the U.S. Patent and Trademark Office is likely to reject my application. However, if I take a clipping and purify, package, intensify, or in other ways make a product *derived* from that clipping, I may be granted a patent alongside the thousands of other medical-device or pharmaceutical patents. It was obvious to the office that the DNA and RNA that had been "purified" in some way merited patent issuance. What must be answered is whether or not the process of describing disease-related genes is more like examining a "natural form" (like the regular clipping from the tree) or more like changing nature to serve a novel purpose (like making a special tea from my clipping).

There is little hope of demonstrating that a disease-related gene mutation is natural, out there to be discovered and possessed of a priori identity. Disease genes are identified in the application by their phenotypic products, when some bits of DNA can be put to explanatory use for some diagnostic purpose. But this is true also of Nebraska. Nebraska is a state only when we say it is. Its geography, climate, and flora are identifiable only after we establish the boundaries for some other purpose. We do not patent Nebraska. Patents for disease-related gene mutations and genes are by contrast useful when the innovation involved in creating some genetic diagnosis product is useful, novel, and nonobvious. While finding a new gene requires no new or novel piece of equipment and involves no "purification" by probes or other "artificial" tools, the work of identifying the group of people who possess a phenotype, the specific methods by which mutations are associated with a particular phenotype, and the methods for putting the epidemiological evidence to specific work in making a diagnosis are clearly synthetic and novel and are not themselves natural phenomena.

Having a phenotype, like having a disease, is in part a matter of social and scientific convention about which states of human life possess relevant or important differences meriting medical intervention or classification. Clinicians and patent clerks sometimes forget that even when a gene is highly correlated with a particular disease, that doesn't make the disease "genetic." Finding DNA is a discovery. Correlating it with human life for the purpose of creating a diagnostic process is innovation. This is the kind of invention that characterizes the world of patents on software more generally—and the patents of disease-related genes are indeed a kind of genomic software program, a part of a new era in intellectual property rather than some sort of challenge to the common property of all humankind.

Thus not to allow patent protection for disease-related gene work on grounds of the meaning of "discovery" and things "natural" seems misguided. If making an extract of my tree clipping can be classed as innovative and patentable, why should we allow opponents of patenting to persuade us that the line between discovering nature and making technology is unblurred for work involving genetic diagnosis? It is in fact a blurry line, and the best we can hope for is to limit the use of patenting to cases where it clearly protects the making of particular diagnostic tests and elements of viral vectors for gene therapy. The U.S. Patent and Trademark Office does not have special access to the heavens, and to expect that it will treat all disease-related gene mutations and genes as "purely natural" phenomena is to ask them to buy into a particularly intractable element of genetic essentialism: the claim that genes are natural things, things "in themselves." So far the patent office has wisely refused to endorse this analysis, opting instead to patent genes when they are employed in a novel product even if that product is tied directly to embodied human biology.

Think about the matter in intellectual rather than simple physical terms, and things seem clearer. When the anthropologist begins to observe a new community, her work is quite clearly that of the observer, and her work products are most clearly associated with existing and well-recognized anthropological methods. She may publish initial findings about the community, but she will not yet be doing synthetic work. She is pointing a telescope into the heavens, cataloguing a new star. As her work proceeds, though, she will increasingly develop new ways of understanding the community and new modes for gathering data. Much of her new data will take on new forms and be published in different ways, for example, as personal reports from her slow, partial assimilation into the culture.

Let us imagine that our anthropologist is approached by a U.S. soft-drink company for advice in understanding this new market. Using her synthetic work, she may develop for them a scheme for interpreting the wishes of the community, using an instrument that will tell who will want to buy soda on the basis of signs and signals otherwise uninterpretable. The U.S. Patent and Trademark Office will be receptive to her intellectual patent application for the methods for detecting such receptiveness, even though she began with simple and well-understood methods for gathering knowledge. Her late turn to identify a way of thinking about the community for the purpose of "diagnosing" its interest in soda is novel and, as best I can tell, patentable. My point by analogy is that the teams working on genetic diagnosis are entitled to the same protection for whatever diagnostic utility they derive. However Faustian is the anthropologist's appropriation of her early project, it would be incoherent to insist that she is merely patenting nature. She, and the patent applicants for gene patents, should be able to request patent protection for innovative scholarly work that mutates and culminates in novel products.

The Dilemma of Ownership

The debate about whether or not genes for genetic tests are patentable would not be as interesting, in fact would not be much more than another in a long line of bizarre puzzles for law professors and jurists, were it not for the second and more important argument against gene patents. This argument is the one about which there seems to be an emerging consensus: the current process of patenting genes, however constitutional or coherent, will lead to erosion of the most basic values of science and perhaps bring harm to patients.

The evidence is pretty compelling. Many university labs have been sued on the grounds that their testing programs, although typically not for profit, are engaged in "commercial" testing and thus infringe on the patents of those who control the genetic test in question. Entire programs have been shut down at the best universities in the nation, and thus has been lost the potential for regional research about patented mutations, the potential to employ university-based genetic counselors in such programs, and the sense that genetics is an activity in which everyone is free to test everyone else's conclusions. On the horizon is a transition from molecular pathology as a research activity in which universities participate for the purpose of research and clinical activity, to molecular genetics as a business. The rules for research, as we have seen, are different for biotechnology conducted outside the university. And, because few rules exist but those of the market, the incentives to cut corners in research will be, and already are, much more acute in private genetics and genomics.

The real danger of patenting in genetics is that those who own the "franchise" of a genetic association are in a position to regulate clinical care in dangerous and powerful new ways. Should Myriad make the decision about whether or not couples test their fetus for BRCA-1 mutations, or should that be a more communal decision? When a physician is awarded a patent on the strength of his research on patients, many of whom would later need the test, should that physician be able to collect royalties against his monopolistic hold on the gene, or isn't that exactly the sort of conflict of interest that led to rules prohibiting physicians from referring patients to facilities in which they have a controlling financial interest?

What is striking is the extent to which these problems lie fallow. Patents have barely been discussed in the public arena, let alone framed as a problem for our schools and courts and legislature to solve. This is a practical problem that requires practical measures, and governmental action should limit patenting so that it does not tie up research, including clinical testing, and does not extend for such a long time. Gene patents should be granted for at most three years' time. More, those who oppose patents should follow the imaginative example of Francis Collins, who, running for his life after the program at Celera Genomics succeeded in all the ways he had promised it would not, urged and assisted in posting as much information on the Web as is found in the federal (or some corporate) efforts, so that the patenting of much of the genetic information about humans will simply not be possible. Clinicians must be disciplined when their commercial activities violate the basic codes of conduct governing medical practice, as has been suggested by Mildred Cho and Jon Merz, who have also documented the effects of a lack of such discipline on patients. But first and foremost, the public must be brought into a conversation, in town halls, over the Web, and in novel experiments in governance at the local and regional levels.

What Can I Do?

When a genetic test is a clinical option, the first question that comes to a patient's mind is not "who owns this?" But perhaps it should be. Deciding whether or not to use genetic services is not like choosing a doctor or electing to have surgery. It is much more like choosing a brand-name drug over a generic. In the coming decade, genetic tests that are not much more predictive of your future (and in many cases are less predictive) than a paper-and-pencil analysis of the patterns of disease inheritance in your family will be offered at top dollar. A genetic test might make sense for you in some cases—even if it is not yet refined, and even if taking the test might impact your and your family's insurability or state of mind. But before you pay top dollar for that test, ask yourself whether you need the information now or would be better off waiting until the "generic" becomes available, that is, until the patent runs out. Sure, you could sue the companies that claim to own the exclusive right to look at your genes, and at some point such a lawsuit may well wend its way to the highest court in a major nation and unravel the web of patents on genes and genetic tests. But if you are not ready to fight that fight—and until now few patients have even considered such a fight—it is worth asking whether or not paying for a genetic patent by taking the test now is really such a great investment in your health—particularly if you are testing for susceptibilities that are years in the future. Karen might be better off without a test, and genetic counseling might tell her just as much, if not more, about whether or not she needs to make serious decisions about what to do to prevent or cope with a future that includes cancer and other diseases.

geneware and the
new infertility

prior to the twentieth century, infertility was a
peril rather than a disease. Those who could not have
children but were of childbearing age have mourned
for centuries, in some of Western cultures' most
important stories, the inability to have a family or to
give someone a child. The fact that human females
have historically been fertile for perhaps 30 percent of
their lives, coupled with the nature of fertility (requiring two people, and thus a relationship), made it difficult to think of the absence of children or the inability
to make a child as a disease.

Twentieth-century birth control and infertility
treatment changed everything, revolutionizing how
sex and relationships would be perceived. By the time
the Genome Project began, parents routinely spoke
not only of infertility but of wanting to pass on genes,

and parents who involved third parties (sperm donors and egg donors, for example) lamented the fact that the child would not have all of the genes of the family. Laws about inheritance, which had been written in a time when there was but one way to have a child, were stretched by questions about who owns frozen embryos and who counts as a father or mother in multiparty assisted reproduction.

A desire has begun to emerge in the twenty-first century among virtually all infertile parents to think of their situation as a disease, that of being unable to pass genes. Today infertility often means a struggle to get your genes, or at least good genes, into a child. But at what cost?

Couples visit the assisted reproductive wizards all around the world, with the exception of the few nations where assisted reproductive technology is illegal. On offer are technologies as varied as sperm and egg donation, the participation of a surrogate mother (who donates an egg and carries the fetus) or gestational carrier (who receives an egg from someone else and carries the fetus), the use of *in vitro* fertilization by postmenopausal women, the mixing of parts of eggs of multiple women, the deliberate creation of twins, sex selection, and more. But what are the trade-offs?

Assisted reproductive technologies are expensive and only partially—at best—covered by public and private insurance in most Western nations. There is some data to suggest that reproductive technologies can cause some kinds of birth defects: children born from *in vitro* fertilization are more likely to be born prematurely, and three international studies suggest that children born from the artificial injection of sperm into the egg may experience developmental delays, though two more studies repudiate this claim. Virtually no study has been conducted on the effect of these technologies on family dynamics or on the psychological well-being of the children. With data that suggest profound psychological effects of adoption on children and families, it is a safe guess that there will be some measurable differences between families made through sex and those made through other, more technological means. But what will the difference be? And what will be the impact of genomics on the high-technology family?

Genomics impacts the way we understand human capacity, meaning, and potential. It is also intimately tied to procreation, sexuality, and reproduction, which are also the foci of the most intimate and invasive social institutions: the family, medicine, and religion.

When humans make children and when it is time to think of inheritance, one is building one's personal and communal understandings of loyalty, privacy, happiness, and growth. Kids of my generation were raised with a story about what it meant to be a child. The idea was that parents loved each other, got married, made love, and babies resulted. Parents loved each other so much that they raised those children as their own, and made sure that they could handle the responsibilities of parenting, marriage, and career by organizing life in such a way that only one of the parents would work, while the other raised the children. It is the story of the birds and the bees. Birds and bees, of course, do not live that way. But the story has powerful resonance for many Americans, representing what has taken on the name "traditional family values" in political discourse, despite the fact that traditional families are an endangered if not extinct species. It is a story that links sex, reproduction, and family in strict terms. While technologies for making children have changed quite a bit, most aim at and are measured against the story of birds and bees.[6] In divorce and adoption, for example, the model of the birds and the bees is used by jurists to measure degrees of variation from the norm, and to aim at giving every child some approximation of the norm.

But the model has aged. If it ever thrived, the *Leave It to Beaver* family is fading into the last millennium. Ward, June, Wally, and Beaver Cleaver and their functional, "nuclear" family are being replaced by dozens of other kinds of families. The political football "family values" is also being replaced, as society gets used to a world of divorce, changing working conditions, and new kinds of reproduction. To be sure, some artifacts of the Beaver model persist. My mother, and perhaps yours, is still sanguine about the ideals of her parents and her parents' parents—ideals of the twentieth century. But in a new millennium, the idea of family is changing fast. Shortly before his death, Benjamin

Spock struggled with the task of rewriting his classic book about child care, noting "I don't know who the audience is anymore."

Family courts, clergy, employers, and society at large struggle to make sense of the myriad new kinds of families. Divorce creates a custody battle for frozen embryos, or the thawing of abandoned frozen embryos brings a request from the Vatican to place, gestate, and adopt—wholesale—thousands of little potential orphans. Single adults, homosexual couples, and couples in their fifties request adoption, ask friends to donate gametes, or arrange for a gestational carrier. Recently widowed women, sisters, or mothers request that children be made from sperm harvested postmortem from their recently deceased male relatives. Couples who use assisted reproduction offer their extra embryos to other couples through "adoption" arrangements, and college students sell their gametes on the open market, through Web pages, making children they hope never to meet and plan to exclude from their estate.

Lots of people still have sex, get married, buy houses, and settle in quaint suburbs. But half of them are eventually divorced, turning then to lawyers and courts to help figure out how to make of their new lives a "family." And a majority use day care more than ten hours each week. So, eventually, whether through adoption, infertility treatment, or divorce, most children of our era will have more than one set of "parents." And for even the most ordinary pregnancies a flood of new choices have changed our potential to shape a baby so as to prevent deadly hereditary disease or to enhance the child's future.

When I say that we have begun making genomic babies, I mean that you and I are inventing new kinds of families using new values that come from a newly acquired power to question twentieth-century Western worship of the Cleaver family. Every month sees the announcement of a new method for finding, having, and making a baby. Genomic families will need new family stories.

The Ants and the Termites

The data are fairly clear that tomorrow's children will not be raised in the world of the birds and the bees. Perhaps the most apt zoological metaphor for parenthood in this time is that of the ants and the termites, which live in large groups with distributed parental roles. The twenty-first-century American culture sees children most often raised by some combination of nongenetic parents, or by those who are not parents at all. More than 40 percent of those born after 1998, we now believe, will have more than one mother or father by age eighteen. The majority of American children are effectively raised in day care, while all three or four of their parents work. Many in society have held that a critical role one can play in the life of a child is that of godparent, coach, or foster parent, and many families in many ethnicities have well-articulated roles for these mentors. It is not accidental that many of these roles have for centuries been identified as parental in nature, despite the lack of genetic or biological connection of the adult to the child. Such is the case for ants and termites, who distribute the parental role seamlessly across many kinds of caretakers, most of whom have no literal gestational or fertilization link to the young. The model of the ants and the termites seems quite contrary to the sociobiological model of modern human reproduction proffered most prominently by Richard Dawkins's *Selfish Gene* model, in which all beings—and each being within the kind—seek to be parents by trying desperately to give genes to someone through sexual reproduction.

The Racehorse

Studies indicate that the vast majority of parents do not disclose the use of donor gametes to their children.[7] In fact, a retrospective study of parents who conceived using donor insemination found that 86.5 percent had not told the child and did not plan to tell and 40 percent had told no one at all.[8] The reasons for nondisclosure were that it would complicate the child's life unnecessarily; that the social parent is the real parent as far as they are concerned, so disclosure is unnecessary; that it could hurt the parent/child relationship; and that the couple had intercourse after the insemination so the husband could be the biological father. Though these reasons appear compelling, it is interesting to note that all save the last are reasons that could be given for nondisclosure in the case of adoption (as well as one form of surrogate motherhood).

Yet, at long last, years into the process of adoption, it has finally been recognized that children should be told by their parents as early as possible that the family was created through adoption. The reasons that this should be true for children of egg and sperm donation are fairly obvious and compelling. Children who do not know may seek or fail to seek treatment because of their false health history. Data suggest that keeping a lie of this magnitude in the family creates tension in the family. And because most parents tell someone other than the child, and due to increasing use of genetic testing, the child will most likely find out anyway and at a more inopportune time. Most important, children might reasonably be said to have some interest in and entitlement to know about what sort of decision their parents made in having them. They might even be said to have a right to know about their genes.

Children of sperm and egg donors will need a story. How can one explain the insistence on donor sperm? How does one explain the selection of a particular donor, or the concealment of that person's identity? Perhaps these children will be told a story about the racehorse, bred from chosen samples. The story isn't sensitive, though, and as one investigates choices about donor gametes it becomes clear that there is a sense in which parents feel that they are "giving" a child something in lieu of one's own gametes.

The Litter

Not every couple will receive a free minivan for having multiple babies in one cycle. Few receive a free house. But the McCaughey septuplets were a miracle, or so the world was told when they were born in a small community in Iowa shortly before the Thanksgiving holiday in 1997. The McCaughey family really wanted a child who was related to them by biology—who shared their DNA—and believed God wanted that too.

In a year that had already proven that reproduction makes for good entertainment, the septuplets cemented the status of children, parents, and scientists in the evolving late-twentieth-century neo-news media, in which a narrative about a castaway Cuban child, a Chilean adoptee, or a mother seeking to hire a surrogate to bear her dead son's child can be told as both news and as an inexpensively produced tale for evening television. It is a story about how amazing is the quest to make more people who share your DNA, and one that elevates the quest to the level of religion.

Seven children bounce around the McCaughey household these days, or so one gathers from the endless coverage of their birth and growth. If all seven are not fully recovered from the circumstances of their birth, by all accounts they seem to be happy and well fed. Ushered into the world by genomics, the McCaugheys' unusual family of eight children—one born before the septuplets—is both a miracle and a puzzle.

The McCaughey family truly did enjoy a miracle. Never before in modern history have so many done so much for a family. A free house, a van, and an unlimited supply of baby food greeted them as they arrived with their kids. Miraculously, they elected not to sue the clinical team that failed to properly monitor their fertility treatments so that they could have easily avoided the expensive and profound situation into which they fell. Understandably, the hospital appears to have cut them a break on the astronomical costs of bringing their children into the world. Church members, relatives, neighbors, and others have gone far beyond the call of duty for day care, story time, and diaper duty. The miracle isn't that fertility medications caused multiple births--that much was predictable. The miracle is that a community, a state, and a nation rallied around this unusual family to help give these kids a happy and healthy life.

The puzzle is what to learn from the septuplets, and what their story will be. Physicians and scientists tell us that the odds against the successful delivery of seven babies are still more than one thousand to one. If couples request strong doses of fertility drugs with the hope of having five, six, or even seven babies, what ought physicians to say? Lots of families have five or six children, but very few have had them all at once. The birth of septuplets cost the community millions of dollars because of the infants' low birth-weight and complex delivery. If even one of the babies had been critically ill, the cost could have tripled, as happened in the Houston case of octuplets.

The costs of multiple births, even below six, are sky-high. While raising large families, sometimes in close age proximity, is very common and age-old, the delivery of all of one's children at once involves extra costs in the short and long term. After the birth of the McCaughey babies, families with four or five multiples asked, "What about us?" Can the community be expected to always provide special resources for every family of assisted reproduction? Or should couples be required to put down some kind of retainer to guarantee that the costs of their procedure will not fall to the community? If a family elects to pursue an expensive reproductive course, who must bear the long-term cost? Could physicians require patients to agree in advance to reduce the number of multiples, by selective termination, to less than three as a precondition of offering these drugs?

What about couples who want to have the whole family at once? Technology to make septuplets safe may be a long time coming, but enhanced safety for multiples below seven may not be. Within a decade couples who want twins—even identical twins—will have new options and the mechanisms to allow those options to be part of the ordinary world of parenthood and childhood may follow.

It is a question so difficult that all seven McCaughey children may not be able to develop an answer, though one can bet that they will be asked to do so. And in a way that is the point. New technologies necessitate new stories. Octuplets and septuplets will be the first in the human species to hear a story of the dogs and the cats; about being part of a litter. Is that a fair story? Is it possible to flourish in a litter of humans? Orphanages house many children of the same age, but with massive staff and under different circumstances. Over time many have proposed the collective raising of children all at a time. There might well be advantages for families and specifically for children in such arrangements. Yet it remains to be determined how, by whom, and where a story will be developed for a child whose entire first-grade class and soccer team comprises siblings. Genomics made possible a redistribution of the children in the community—in a way no one in the Genome Project would ever have imagined.

Choosing to Be a Mother Late in Life

When Arceli Keh brought her fake ID to California, it was her last attempt to find someone who would let her have a baby. Nurses at other institutions hadn't believed the driver's license was real because Keh looked older than the age, fifty-three, it gave. But as she would tell the entire world after she found a clinician who either believed or overlooked the ID, her desire to have a child at the age of sixty-three was predicated on her need for and love of children. She was strong and so was her older husband. She had others in the family. When the world said of her case, "How odd!" she was silent, rocking that child.

The world asked the predictable questions. One histrionic French bioethics scholar told the couple that it was irresponsible to make children after menopause because "the right to procreate must stop where nature intended." Meanwhile, men ninety-five and older continue to marry and father children.

The technology that gave Arceli Keh a child has since been used to provide babies to many women at or past the age of menopause. And equally well developed are new technologies to give young women the power to wait—to delay pregnancy as long as two or three decades—through the freezing of eggs. Age will no longer be a barrier to motherhood.

How many women would want to use the technology? How many would be willing to try to have a child after menopause, to trade some risk of minor developmental delays in offspring or of miscarriage against the known realities of having children early in life? Recent arguments to the effect that having kids early makes more sense for women in medical terms do not take into account the future of advances in genetics and reproductive medicine. Within a decade the risks that today motivate women to abandon the most important years of their careers to childbearing may be a distant memory—but in the meantime, the decision to have children late is and has long been laced with uncertainty and misgivings. To make matters worse, arguments on both sides of the "should I have kids early" debate feature images of good and bad mothers, good and bad women, rather than the importance of choices in reproductive life and family.

The argument against late pregnancy or at least against *planned* late pregnancy is that it seems, to many scholars, to challenge what is natural, what is genetically normal, or at a minimum what is typical for women. The arguments in favor have focused on simple reproductive freedom as a high human aspiration, but have also done so at the expense of women. Women who choose to have children earlier in life are left wondering whether the ability to put off having a child will end up putting additional pressure on twenty-four-to-thirty-four-year-old women to use the high-tech way to have a child, by decreasing the incentive of employers to offer real benefits to working moms in the prime of their careers.

So why not have a baby at sixty-five, fifty-five, forty-five? The arguments against late gestation range from the sublime to the political but none really offer the kind of practical help that women in their twenties, planning careers, or in their fifties, planning the latter half of their lives, are looking for. Building a family later in life means having a story to tell yourself, your family, and especially the children about what it means for you to be slightly older and to have different experiences from other moms. It might mean committing to a regimen of self-examination, and it definitely means building a community with whom you can talk about the special needs that come with building such a family. Arceli Keh's mother, believe it or not, as well as her elderly husband and extended family, all pitched in to help with the baby and with the whole idea of her carrying a baby to term. If you don't have a support system, building an unusual kind of family will be especially challenging, and there are no secret solutions to that problem.

The Lions

Everyone has a mother. Some children have biological mothers, some have stepmothers, some live with adopted moms, and some have moms who donated eggs or carried them as embryos for other moms who would raise them. Some kids have moms whom they know well, others have moms they don't know, and still others have moms they don't know exist. Many have godmothers. But what if you had three or four mothers, each of whom was genetically linked to you?

Some children grow up without the presence of a mother. Some children grow up and find a long-lost biological mother, and for some of these there is a sense of having reunited while for others there is a sense of invasion and fear. One day children may grow up and find long-lost egg-donor moms. One day children may have many moms—a mom who donates an egg, a mom who carries the embryo, and a mom who raises the child.

It is anyone's guess who the "real" mom is in the twentieth century. Nature and nurture both contribute to parenthood, and in a complicated world of new reproductive technology, there are more and more ways to be a parent.

Recently, Jamie Grifo announced that he has added a new mother to the mix: the egg-shell mother. Using a technique similar to that in Ian Wilmut's cloning of Dolly the sheep, Grifo transfers the DNA from one woman's egg into an egg that has been given by a donor and from which the DNA has been removed. He then fuses the new egg together, and fertilizes it with sperm of the first woman's choosing. The woman who donates the "emptied" egg, into which the DNA of another woman is placed, gives the child mitochondrial DNA, as well as some other genetic information that is floating about. The bulk of the genetic information in the new egg, though, comes from the "DNA donor," who will likely be the mom that raises the child. The procedure was not tested in animals or reviewed by the government (a problem we have discussed and to which we will return), yet human babies have been birthed.

Within a decade the mixing of eggs will be regular, an option open to you at most clinics that perform *in vitro* fertilization, and a pool of donors of egg "parts" will be available on the Internet or its equivalent.

Why would institutions want to create a way to make a child with two genetic mothers? Until today, women whose eggs are damaged or who are lesbian have had only a few options: have no children, adopt, or use a donor's eggs. In none of these cases would the genetic information from the woman find its way into any children raised by that woman.

This is troubling for couples because many feel that the use of donor eggs or sperm is a kind of invasion of their intimate life; they planned to make children through sex, and using someone else's eggs or sperm can almost feel like adultery. Other couples just want to continue their genetic bloodline, and feel that as long as they are spending so much money on assisted reproduction, they may as well get as close to a "natural" birth as possible.

For men, the use of donor sperm has all but been replaced by the technology of intracytoplasmic sperm injection, or ICSI, which takes sperm that cannot swim or penetrate the egg and injects them directly into the egg with a tiny needle. This lets them make a genetic contribution. Through Grifo's technology, women with mitochondrial diseases and other problems in their eggs will now have the same opportunity, if they can afford it.

There is no way to tell what will happen with the technology. While Grifo's group got approval from a local ethics board, no one has announced plans to study this new kind of family, or to look at the long-term safety effects of the technology for kids. Some ICSI studies suggest that there is a possibility that children made through that technology will either inherit their genetic father's infertility problem or have a new problem in intellectual development.

In addition, there simply is no effort under way to study the medical or social implications of this technology for society. I wouldn't want to be the first child to grow up with the "shadow" of a mom, who contributed something to me but we don't know what that something is. Should I meet her? Does she owe me anything? What if her eggshell contains a predisposition to disease—can I sue her?

Lions represent a story for children who are gestated by one woman, with an egg from another and DNA from a third. Lions interchange the role of mother, and their participation in an unselfish act aimed at aiding one another and the whole is an especially interesting antidote to the selfish-gene model. Children with two moms are at least owed the attempt to find some similar narrative into which that mixture might be fit as a recipe for flourishing.

The Phoenix

Brandalyn was born to Gaby Vernoff, who loved her husband, Bruce, very, very much. The Vernoffs seemed to lose their chance to make a baby with Bruce's tragic death. But

Gaby found a physician, a very special kind called an andrologist, who was willing to take the sperm from her husband's dead body and freeze it in nitrogen. Gaby believed that in her husband's *genome* was the key to completing the act of love that was unfinished when he died.

Gaby wanted to be impregnated by Bruce's sperm to make a baby right then and there, and asked her in-laws for their blessing. The elder Vernoffs asked her to wait, and grieve, before she made decisions about having children, which she did. Everyone wanted to do the right thing. Still, while there is every reason to believe that Bruce wanted to have children, there is no evidence he would—or would not—have wanted to have a baby made after his death from parts of his dead body.

Recent research by colleagues of mine suggested that more than three hundred women have made this request of physicians after the death of their loved ones in recent years, and that was before their study—and the case of the Vernoffs—made its way to the national media. It is difficult to see how one would turn down a wife's request to dispose of her husband's body in any way she sees fit. If a wife wants to cremate the body of her husband, we oblige her. If she wants to take sperm, institutions seem obliged to honor her request. Taking the sperm would not be acceptable if the woman were a girlfriend or just a passerby, and it would not be acceptable to take sperm if the husband specifically forbade it in his will.

But taking the sperm is one thing, and using it is something entirely different. The physician involved in the postmortem-sperm case, Dr. Cappy Rothman, is a world-renowned and kind scientist and clinician known for his warmth and caring. He says, and there is every reason to believe, that Gaby Vernoff and her family had thought long and hard about her desire to make a baby in this way, and that this was in Dr. Rothman's opinion the first case in which a postmortem sperm transfer seemed acceptable. So without describing the activity as research or seeking the approval (not required but often sought as a safeguard) of the Centers for Disease Control program on reproduction, the U.S. Food and Drug Administration, or any of a number of professional organizations with a stake in the matter, he took the next, fateful step and helped Gaby to become pregnant. As he told the story in a debate with me on the ABC News program 20/20 shortly after the birth of Gaby's child, he was not worried about the fact that there was no specific paternal consent, because in his view the Vernoffs were a couple who really loved each other and their child would be the product of that love.

There are overarching moral issues here that we have discussed in a different context and to which we will return: it is wrong to force someone to reproduce, as is clear in the rules about rape and forbidding misbehavior involving the use of cadavers. It is fine for dying men to make sperm donations and for their wives to use them. It is fine to put it in your will that you want a child after you die. But no one is obligated to carry that child for you, and you ought not make that child after your death unless you really did intend to have it.

If we don't have rules like this, before long we will be cloning Abraham Lincoln, Wolfgang Amadeus Mozart, and just about every dead athlete who broke a record and seems to have "wanted children at one time or another." Controlling the dead body of the husband is one thing, but making a baby from it is another.

Think of the child as well. Brandalyn may not be called Frankenstein or the Phoenix, but she will certainly be in line for some counseling. The question for infertility specialists is about responsible practice, and what it means to have a "good candidate" for a fertility procedure. A good candidate for postmortem sperm retrieval is one who gave consent. Period.

And even after consent, there should be some examination of the issues for each family. Will making this baby mean the wife can never remarry? Will it prolong the grieving process or even prolong denial of the death? How long should the family wait? Obviously no maverick doctor should be making this choice alone. Brandalyn is a baby with a very special future and a very unusual past. You wish her the best. But we ought not open the graveyard gates to *in vitro* fertilization until we have better rules.

It may also be possible to make a human baby with an

egg from another species. Recent experiments suggest not only that eggs vary little across different species, but also that human DNA can reprogram the eggs of other species. Advanced Cell Technology of Massachusetts made news when it created a variety of embryolike creatures that consisted of human DNA within a cow's egg. Could such a creature be gestated and birthed as a human? Probably, although probably not with a cow's egg. As transgenic egg donation from primates or other mammals finds its way into human reproduction, stories for that technology too will be needed. Perhaps we will find these stories in reruns of *The X-Files*, or perhaps this too can be a "normal" way to make a family. It is all in the language and the story that language creates.

Your Story Is My Story?

Perhaps the most discussed form of new family to come from genomics is the family of a human clone. Cloning is an astonishing technology that literally moves a genome from one person into a future person—or at least that is the plan. Human somatic-cell nuclear transfer, otherwise known (somewhat inaccurately) as creating an embryo by "cloning," involves the starvation and subsequent implantation of DNA from specialized, nonsexual cells of one organism (e.g., cells specialized to make that organism's hair or milk) into an egg whose DNA nucleus has been removed. The resulting egg and nucleus are shocked or chemically treated so that the egg begins to behave as though fertilization has occurred, resulting in the beginning of embryonic development of a second organism containing the entire genetic code of the first organism.

Mammalian cloning, through this nuclear transfer process, has resulted in the birth of hundreds of organisms to date. However, significantly more nuclear-transfer-generated embryos fail during pregnancy than would fail in sexual reproduction, and a substantial majority of cloned animals who have survived to birth have had some significant birth defect.

Reproduction, or, perhaps more accurately, replication of an organism's genomic identity, does not normally occur in mammals, with the exception of twinning, which always results in the simultaneous birth of siblings. Only plants reproduce through replication from one generation to another. The prospect of such replication for humans has resulted in the most controversial debate about reproduction ever to be taken up in Western civilization.

Part of the issue about cloning is the danger involved in making a clone, danger to the woman who gestates the clone, to the clone itself, and to the social institutions already creaking under the strain of virtually unregulated reproductive technology. But if cloning *were* safe and effective, what story can one tell a clone? Already I have noted that human cloning is unprecedented in the natural history of mammals. Twins are the closest existing phenomenon, and unlike the clone they are born together and have sibling relationships. The media stories of parental roles in cloning are frightening in almost all cases. One has parents replicating a child who has died early due to an accident. Another has an infertile woman seeking a genetic link to her recently deceased husband through a clone from a tissue sample she happens to have lying around. Still a third has the parent raising a clone of his wife to realize his dream of seeing his wife as a child. It is difficult to imagine how a family would form stories for such a mode of intimacy, birth, and connectedness.

Families struggle with new technologies to restore the apparent equilibrium of the "classical" family, and work to find technologies that have as much explanatory power as the birds and the bees. This is one reason why, for example, most couples will use sperm injection rather than donor sperm. It is simply assumed that it is better and more normal to have a child that shares more identity with me. Thinking about and emphasizing the role of children's stories helps to bring these two issues into focus.

Habits in making families are only part of the culture of reproduction. Parenthood is, at its edges, controlled and defined by the community and its institutions, and it is more than idle Platonic fantasy that children are in some sense raised by the state. Economics, politics, and theology play roles in how infertility is understood and treated. The family is also only one among many institutions that raises children. In fact, when parents fail in a variety of tasks (from immunization to feeding to education), they can lose their parental rights, to be restored only at the discretion of representatives of democracy. The upstream manifestation of this public concern for the welfare of children is apparent when, for example, it is argued that future children ought not be exposed to the danger of cloning, or that research to clone humans is of a comparatively low priority in the existing array of choices for research spending. Even editors of scientific journals and newspapers have a choice about what they will send out for review and in what way they will publish findings about cloning. The culture has numerous options as its institutions are reconstructed by the rush to create and manage new technologies for parents and children. One is not limited by the concepts of family values or parenthood from the last thirty years, but neither can one invent ideas of familial rights without situating them in their cultural context.

The Big Question

What is infertility? Is it a terrible disease that plagues those whose gametes or organs malfunction? Is it a disability? What counts as a cure, and what sort of treatments do the infertile have a right to expect? It is the ultimate question for a world in which families are becoming genomic: do the genes make the child?

The Question of Sameness

Is the definition of a "normal" child one made from two parents' DNA? Is it a disease to lack the ability to make that kind of child? I am not sure. One thing that is clear is that the twenty-first-century framework within which such children are the benchmark of normalcy has led to a tendency to put too much emphasis on having kids who are similar to us, a desire often attributed to those who seek cloning but that in fact is implicit in virtually every expensive reproductive technology. There is a big political and social question here that must be answered by every potential parent: is the desire for children whose traits complement those of the parent morally superior to the pursuit of sameness? Are technologies that look like they are restoring patients to a natural state—the state of being able to have a normal child—really just a primitive way of thinking about family and genomics?

In the new genomic family there are two choices. One is the development of assisted reproduction aimed at making families more alike. The other is the development of a new model of the family.

We can all understand the longing for children. Reproduction is a matter of urgency, excitement, or even trepidation for many of us; it accounts for much of our motivation for mating, marriage, and family. The failure of reproductive organs to function effectively, preventing reproduction through sexual intercourse, is a painful experience for many patients of reproductive medicine. Once diagnosed, infertility turns your private sex into a public matter because it takes more than two people to utilize infertility treatment, assisted reproductive technologies, nuclear transplantation, or adoption. Children of adoption and infertility treatments also sometimes return to their agencies or donors "of origin" for further information or assistance, adding still more people to the mix. Some have argued that it is unfair to further challenge infertile individuals by requiring them to meet special legal standards, such as screening for "fitness to parent," that are not applied to fertile couples.

At the heart of claims about a right to infertility treatment and advanced reproductive services is the idea that there is something valuable about having the *kind* of child obtained through sexual reproduction. Children of sexual intercourse share chromosomal material with their parents, are gestated by a parent, and thus are participants in the many institutions and rituals that revolve around being "like" a parent. Johnny has his father's eyes, his mother's smile, even Aunt Edna's sense of humor. These seemingly insignificant similarities are indeed crucial to families' self-understanding and create for many families a wellspring of identity and commonality.

For many patients of reproductive medicine, assisted reproduction can be a way to participate in the experience of commonality or can increase the commonality experience (e.g., couples utilize donor sperm as opposed to adopting a "totally unrelated" child). The question is how meritorious is the goal of producing a genetic relative, independent of whether or not it is the most frequent goal of Western parents?

The right to similarity has never been debated among scholars of reproduction and bioethics. So, rights, as Ronald Dworkin quipped, grow on trees in Western jurisprudence. John Robertson, the leading defender of reproductive liberty, assumes, without arguing for, the positive right of infertile parents to pursue a child whose similarity to his or her parents is programmed into the genes. For him, the pursuit of similarity is just part of the "natural" process that can be taken for granted by fertile couples, a process that should to the greatest extent possible be restored for the infertile.

But the pursuit of a child need not necessarily entail a pursuit of perfection or a pursuit of genetic similarity, and when parents seek children that are traits-perfect or identical to themselves, society correctly frowns. In the debate about whether ultra-high-technology parenthood should be allowed or even built in to health insurance, the focus has been on the rights of parents to have *a* child, the safety of new technologies for the fetus, child, and germ line, and the science and politics of research on embryos. What would it mean to say that infertile couples or couples with mitochondrial or other avoidable hereditary disease have a "right" to the pursuit of similarity? Further, what does it mean to say that such a right is *positive*, conveying either the parity requirement that insurers pay for such similarity, or that research be funded to enable the eventual pursuit of similarity by those who can afford to purchase it?

Given the biological consequences of passing along one's genes, which include retaining certain negative traits in the human gene pool, should passing genes be defined in any sense as a basic human right? While I may increase the chances that a child will share certain of my physical and mental features by reproducing "genetically," this in no way demonstrates the societal desirability of doing so. Nor does it, if society looks beyond traditional practice, establish my right to do so. What is missing in the debate about rights in reproduction is an argument for guaranteeing genetic similarity to those who engage in assisted reproduction, a specific claim for what kind of similarity "matters" and why. As the techniques of reproductive medicine are refined so that it is possible to separate the genetic aspects of reproduction from the other aspects—affirmation of love, heightening of sexual intimacy, experience of pregnancy, childbirth, and childrearing—institutions and individuals all around the world will need new analysis about what the family means.

Being Similar and Being Geneware

To claim the right or preference for the pursuit of similarity is to assume the superiority of one kind of community over another. The study of the role of different kinds of families as a model for society has been shaped in large part by the work of Ferdinand Toennies, a turn-of-the-twentieth-century historian and biologist. Toennies makes a distinction between the *Gemeinschaft* community, in which the dominant bond is commonality or sameness, and the *Gesellschaft* community, where the dominant bond is complementary traits and/or mutual interdependence.

The *Gemeinschaft* family wants kids that are like the parents, speaking the same language, often working in the same environment, often named Junior. It is a separatist family, separate in the sense both that the family that shares a language often can communicate effectively only with other family members, and in the sense that the *Gemeinschaft* family often has strong views about what it means to leave the family or bring outsiders into it. Separatism makes a good family and political strategy early in the combination of two communities. Toennies points out that *Gemeinschaft* communities typically begin as a union of the similar, bound together by similarities simply in the hope that a body of similar people will have a better chance in their pursuit of equal treatment, as in the case of African-American separatism in the 1970s or immigrants' insistence on maintenance of their language of origin in the United States since its inception. Separatism can begin as a quest for tolerance or at least the end to some persecution, as in the case of Polish struggles in the 1980s. Each of these efforts fits Toennies's profile of the emergence of a community. It makes sense to band together when there are threats to the family or culture, problems with the loss of power through assimilation, and a strong challenge to the identity of individuals, ways of life, and even the meaning of the family.

However, as a long-term strategy for building family and community, separatism does not work very well, and its emphasis on the singular importance of children who share Mom's and Dad's traits and habits can result in the kind of danger that was forecast in condemnations of human cloning. Children who have no freedom to build their own narrative are literally stuck in their parents' language and dreams in a way that is much more pronounced than for other children in the *Gesellschaft* family and community.

The search for diversity between family members is as vital in the establishment of the parent-child community as is the longing for sameness. Children fail if they only repli-cate their parents' existences or at least so most of Western theology and literature would suggest. The more enriching parental experience occurs when children draw parents into new worlds and thereby enlarge the parents' humanity. This may be most evident as children mature into adult-hood by affirming their own identity. How many parents have been dragged into the odd habits and affinities of their children only to find that in this difficult act they found the meaning of moral and family growth?

In a way this is the kernel of what is often called the American dream, reinscribed in countless narratives of just the sort of ethnic and political communities that began in

Toennies's separatist framework. Parents in America want their children to have more opportunity, more latitude, and more education than they themselves enjoyed. It is ironic then that much of the energy children of the 1950s and 1960s have expended is directed at making new and better communities that are nonetheless stultifying, censoring, and subject to the boorish and jingoistic whims of false history that always accompanies separatism.

Very few communities in the United States really need separatism to flourish, and the ultimate goal of making children who complement us rather than copy us can be a beacon of tolerance in the development of new technologies for reproduction. Assisted reproduction that aims at making sure my kids are really "mine" in the most narrow sense imaginable sets a terrible tone for the parent-child relationship, making reproduction a literal rather than metaphorical description of the role of the parent.

Redefining Infertility

In the dark, often scared and embarrassed, the infertile sometimes suffer less from organic dysfunction than from a faith in bad genetics. Richard Dawkins proffers the most egregious claim: we are selfish organisms, out to make more of ourselves, and the desire to pass on genes is what makes us fit, normal, and happy. Even recent court decisions about infertility embrace a naïve genetics of selfishness, in which all those humans who cannot make children through sex are pictured as disabled.

The inability to have children on demand through sex presents not a disease but a difficult and often very painful range of new choices in the genomic matrix. What will the couple do to satisfy its profoundly human need to nurture and love a child? How will that need be framed for the couple so that each can say that his or her need for family and reciprocity with the young has been fulfilled? For couples who deeply need a child that is similar to themselves, one can only prescribe deep thinking, intense and invasive, to investigate the meanings of that desire and to probe how those meanings might manifest themselves in a child. Even parents who are perfectly fertile but have such desires need to think long and hard today about how much their definitions of family depend on the old views of genetics. Genomics means new choices, which can be liberating or constraining depending on how much you know and how well you understand the software. A child who comes to you cold and homeless, sharing none of your genes and fresh from the world of thousands of unadopted orphans, presents a challenge as healing of infertility as the effort to make one. One profound miracle of the mapping of the genome is that it is now more clear than ever that we share so much of our genes with every human being that to select a child on the basis of a few inherited susceptibilities or traits is to overestimate the power of individual genes to make us human, to make us families, or to link us together.

The key proposal for the era of genomic children is that technologies to transfer similarity from kids to parents, at which my imaginary machine aimed back in high school, should not be viewed as a high research priority. In a culture where thousands of "less desirable" foster children and infants go unparented, and where many poor pregnant women and couples receive no prenatal or postnatal care from physicians or genetic-testing services, it scarcely seems prudent to institutionalize an entitlement (against social resources) to the practice of nuclear transplantation or assisted reproduction.

While more than vanity, the transplantation of our biological identity to children is in its proper role when it is seen as a distant second to health, complementary development, diversity, and other moral goals. Sameness of values is a struggle for every parent already, embodying many of the trade-offs that future parents should discuss when infertility comes up.

It is moreover a mistake to assume that the pursuit of similarity is embedded in the behaviors of parents who sexually reproduce. Indeed the obvious pursuit of diversity in our human mating behaviors, and the diversity of kinds of pursuits in mating, suggests that "normal" reproduction never consists of a simple attempt to copy DNA to offspring. In addition, parents of adopted children and others engaged in socially constructed parenting are frequently reported to "look like" and "act like" their parents; much of the similarity we seek in our children is constructed by the assumptions of society about how family works, rather than by genetic transmission.

Advocates of reproductive liberty are right to insist that cross-examinations of the infertile "to see if they are good enough," overbilling for infertility procedures, and restrictions on infertile couples are a mistaken strategy for moving forward in the genomic era. But assisted reproduction does raise questions about the motives of all who long only to have "one like me," motives that are central in the debate about cloning: why is sameness so important, and how will that desire manifest itself in the care of the child?

As society balances its research budgets and allocates pragmatic resources, the value of research in assisted reproduction will tally far behind other demands on social resources devoted to reproduction. Potential adoptees whose similarity to parents might easily be socially constructed, and whose potential complementarity is significant, should not go unparented simply so the wealthy can pursue parental vanity. All children have a right to, and parents a responsibility to plan for, a supporting environment. No child's family story should be, in itself, a disability.

conclusion

want to be a beta tester?

genetic information in the twenty-first century will
be extraordinary, and the combination of software and
genetic hardware will enable bold new choices.

Genomic endeavors are still, for the most part,
made up of huge institutions whose aim is to gather
data or to place it on maps. But every exponential
increase in computing power launches a thunderous
boost in the world of genetics. Most of the work and
the majority of the workers are running programs
instead of physically manipulating genetic material. In
fact, robotics has taken over the cultivation and
manipulation of the raw materials.

How much longer until the genomics program will
run on *your* laptop? How much longer until you can
identify and analyze genetic information on a hand-
held device? Not long. Powerful new diagnostic tools

that assay key proteins and other molecules are literally creating thousands of new categories of disease, and for every kind of gene-to-malfunction correlation there will be a corresponding gene-to-surplus correlation. The key is software.

A typical computer can easily spit out billions of calculations each second, but you cannot read them on a screen unless they are rendered through symbol systems you can absorb. Fractal geometry is astonishingly complex even at slow speeds of calculation. A stream of fractal information is dizzying. No "normal" person can absorb information or interact in the style and speed of a computer's "native" language, so software makes the computer's stream intelligible. Data becomes fractals, which become a hypnotic screen saver.

For now, genetic technology imitates the personal computer's ideal interface with human users; CD-ROMs teach families about the risks and benefits of a genetic test for breast-cancer susceptibility. Maps of genetic material are gathered into Web-friendly databases.

The challenge of "buying" genetic technology, such as the decision to use a genetic test, is often compared to the challenge of learning a computer language. The contemporary genetics lab is filled with computer technicians, who feed biological stuff to a computer for analysis and compilation. But genetic innovation will not be slowed for long by the clumsy and expensive computers currently used in research and development. Likewise, the end user's interface with genetic information will be much more powerful and interactive. Genetic information will be rendered in graphic form and with an eye to the uses you have at your disposal.

In a few years you will be able to upload information about your genes and get customized information about your health. Home genetic technology will let you choose activities and environments that play to your advantages or challenge you in areas where you have a deficit compared with others in the database.

Home genetic technology will take its cue from the same classical assumptions made by the twenty-something college dropouts who invented personal computers in their garages. The goal is to build software and hardware that is modeled to the needs of a typical user. Most people would like to communicate more effectively, to store and retrieve more information, and to plan and calculate with more intelligence. The mouse, the word processor, and the Web browser each come from biological and ecological models thought to resonate with human habits of mind and body.

If the goal is disease prevention, the software program will take a home genetic sample and identify those risks with which specific mutations you carry are correlated. It will then identify the ways in which you can reduce your risk using behaviors and treatments that have been developed to respond to those with your genes. More, it will identify the kind of nutrition that your body needs, and the sorts of "inputs" that won't work out well for you: smoking, fatty foods, and alcohol might be harmless for a significant portion of the population, and pesticides might cause cancer in another portion. Perhaps you should live close to sea level. These are the kinds of correlations that rely only on advances in software and entrepreneurship; the basic epidemiological science is ready to go.

The hope is that a new generation, schooled in geneware, where ideas are beta tested before they are released as products to the wider community, has begun to work on genetics. The insight that comes from new kinds of analysis of genetic material may pale in comparison with the insight that comes from more than a decade of software engineering.

As genetics benefits from and benefits the digital world, a new way of talking and thinking about genetics, much less traditional and much more promising, will see an entirely new generation unlock an entirely new way of living in which being genomic is not so much a problem as an opportunity.

notes

1. U.S. Constitution, art. 1, sec. 8, cl. 8; *U.S. Code*, title 35, secs. 101, 102 (1982).
2. A. L. Caplan and J. Merz, "Patenting Gene Sequences," *British Medical Journal* 312 (1996): 926; see also P. Halewood, "Law's Bodies: Disembodiment and the Structure of Liberal Property Rights," *Iowa Law Review*, July 1996; A. Kimbrell, *The Human Body Shop: The Engineering and Marketing of Life*, (New York: Free Press, 1993); C. Barrad, "Genetic Information and Property Theory," *Northwestern University Law Review*, 87 Nw, U.L. Rev. 1037.
3. J. Merz, M. Cho, M. Robertson, and D. Leonard, "Disease Gene Patenting Is a Bad Innovation," *Molecular Diagnosis* 2, no. 4 (1997).
4. Ibid.
5. G. McGee, *The Perfect Baby: A Pragmatic Approach to Genetics*. (New York: Rowman & Littlefield, 1997); D. Nelkin and S. Lindee, *The Gene Mystique* (New York: Free Press, 1993).

6. Glenn McGee and Ian Wilmut, "A Model for Regulating Human Cloning," *The Human Cloning Debater,* ed. Glenn McGee, (Berkeley: Berkeley Hills Books, 1998).

7. S. Klock, M. Jacob, and D. Maier, "A Prospective Study of Donor Insemination of Recipients: Secrecy, Privacy, and Disclosure, *Fertility and Sterility* 62, no. 3: 477–84 (1994); see also S. Klock and D. Maier, "Psychological Factors Related to Donor Insemination," *Fertility and Sterility* 56, no. 3: 489–95 (1991).

8. Klock and Maier, "Psychological Factors," 489–95.

suggested further reading

Introduction

Fukuyama, Francis. *Our Posthuman Future*. New York: Picador, 2003.

Keller, Evelyn Fox. *Making Sense of Life: Explaining Biological Development with Models, Metaphors and Machines*. Cambridge: Harvard University Press, 2002.

Negroponte, Nicholas. *Being Digital*. New York: Vintage, 1995.

Nelkin, Dorothy and M. Susan Lindee. *The DNA Mystique: The Gene as a Cultural Icon*. New York: W. H. Freeman, 1995.

Phan, L.; Doukas, D. J.; Fetters, M. "Religious Leaders' Attitudes and Beliefs about Genetics Research and the Human Genome Project," *Journal of Clinical Ethics*, 6(3):237–245, 1995.

Wade, Nicholas. *Life Script*. New York: Simon & Schuster, 2002.

Chapter One: Bits and Genes

Beskow, L. M.; Burke, W.; Merz, J. F., et al.; "Informed consent for population-based research involving genetics," *JAMA* 286(18):2315–21, Nov. 14, 2001.

Buchanan, Allen, Dan Brock, Norman Daniels, Dan Wikler. *From Chance to Choice: Genetics & Justice.* New York: Cambridge University Press, 2000.

Burnham, Terry and Jay Phelan. *Mean Genes: From Sex to Money to Food, Taming Our Primal Instincts.* New York: Penguin, 2001.

Dawkins, Richard. *The Selfish Gene,* 2nd Edition. New York: Oxford University Press, 1989.

Galton, Francis. "Eugenics: Its Definition, Scope, and Aims," *The American Journal of Sociology,* 10(1), July 1904.

Galton, Francis. *Hereditary Genius: An Inquiry into Its Law and Consequences.* London: Macmillan and Co., 1892.

Kass, Leon R. *Toward a More Natural Science: Biology and Human Affairs.* New York: MacMillan, 1995.

Keller, Evelyn Fox. *The Century of the Gene.* Cambridge: Harvard University Press, 2001.

Ridley, Matt. *Genome.* New York: HarperCollins, 2000.

Tagliaferro, Linda, Mark V. Bloom, with a forward by Glenn McGee. *The Complete Idiot's Guide to Decoding Your Genes,* New York: Alpha Books, 2000.

Wolpe, P. R. "Bioethics, the Genome, and the Jewish Body," *Conservative Judaism,* 2002.

Chapter Two: Floppy Genes and Rewritable Genomes

Baxevanis, Andreas and B. F. Francis Ouellette. *Bioinformatics: A Practical Guide to the Analysis of Genes and Proteins.* New York: Wiley, 2001.

Cavalli-Sforza, L., P. Menozzi, A. Piazza. *The History and Geography of Human Genes.* Princeton: Princeton University Press, 1994.

Kay, Lily. *The Molecular Vision of Life: Caltech, the Rockefeller Founda-tion, and the Rise of the New Biology.* New York: Oxford Univer-sity Press, 1992.

Stock, Gregory. *Redesigning Humans: Our Inevitable Genetic Future.* New York: Mariner Books, 2003.

Wu, Cathy H., and Jerry W. McLarty. *Neural Networks and Genome Informatics.* London: Elsevier Health Sciences, 2000.

Zweiger, Gary. *Transducing the Genome: Information, Anarchy, and Revo-lution in the Biomedical Sciences.* New York: McGraw-Hill, 2001.

Chapter Three: Learning to Program Your Genes

Elston, Robert C., ed. *Biostatistical Genetics and Genetic Epidemiology.* New York: Wiley & Sons, 2002.

Frank-Kamenetskii, Maxim D. *Unraveling DNA: The Most Impor-tant Molecule of Life.* New York: Perseus, 1997.

Lewontin, Richard C. *The Triple Helix: Gene, Organism and Environ-ment.* Cambridge: Harvard University Press, 2001.

Parker, Lisa. *Mutating Concepts, Evolving Disciplines: Genetics, Medicine and Society.* Publisher

Sankar, P. and M. K. Cho, "Toward a New Vocabulary of Human Genetic Variation," *Science,* 298: 1337–1338, 2003.

Chapter Four: Bugs in the Geneware: Genetic Testing

Bosk, Charles. *All God's Mistakes: Genetic Counseling in a Pediatric Hospital.* Chicago: University of Chicago Press, 1995.

Doukas, D. J. and J. W. Berg. "Genetic Testing and the Family Covenant," *American Journal of Bioethics,* 1 (3): 2–10, 2001.

Khoury, Muin, et al., eds. *Fundamentals of Genetic Epidemiology.* New York, Oxford University Press, 1993.

Long, Clarissa, ed. *Genetic Testing and the Use of Information.* New York: AEI Press, 1999.

McGee, Glenn. *The Perfect Baby.* New York: Rowman & Littlefield, 2001.

Nightengale, Elena O. and Melissa Goodman. *Before Birth: Prenatal Testing for Genetic Disease.* Cambridge: Harvard University Press, 1990.

Paul, Diane. *Controlling Human Heredity.* New York: Oxford University Press, 1989.

Quaid, Kimberly, et al., eds. *Early Warning: Cases and Ethical Guidance for Presymptomatic Testing in Genetic Diseases.* Indianapolis: Indiana University Press, 1998.

Robertson, John. *Children of Choice.* New York: Oxford University Press, 1994.

Young, Ian. *Introduction to Risk Calculation in Genetic Counseling.* New York: Oxford University Press, 1999.

Chapter Five: Bugs in the Geneware: Gene "Therapy"

Doukas, D. J. "Primary Care and the Human Genome Project: Into the Breach," *Archives of Family Practice,* 2(11): 1179–1183, 1993.

Gillis, Justin. "A Hospital's Conflict of Interest: Patients Weren't Told of Stake in Cancer Drug." *Washington Post,* Section A, p. 1. June 30, 2002.

Nelson, Deborah. "Penn Ends Gene Trial on Humans." *Washington Post,* Section A, p. 1. May 25, 2000.

Walters, Leroy and Julie Gage Palmer. *The Ethics of Human Gene Therapy.* New York: Oxford University Press, 1997.

Weiss, Rick. "FDA Halts Experiments on Genes at University; Probe of Teen's Death Uncovers Deficiencies." *Washington Post,* Section A, p. 1. January 22, 2000.

Chapter Six: Geneware in Your Kitchen

Berry, Wendell. *What Are People For?* Berkeley: North Point Press, 1990.

Mcgee, Daniel B. and Dudley J. Burton. "Genetically Modified Food and Science: A Comprehensive Review." *Journal of Global Environmental Issues*, 3(2):17–59, 2003.

Lambrecht, Bill. *Dinner at the New Gene Café: How Genetic Engineering Is Changing What We Eat, How We Live, and the Global Politics of Food.* New York: St. Martin's, 2001.

Magnus, David and Arthur Caplan. "Food for Thought: The Primacy of the Moral in the GMO Debate." In M. Ruse and D. Castle, eds., *The GMO Debate*, New York: Prometheus, 2002.

Pence, Gregory. *Designer Food.* New York: Rowman & Littlefield, 2002.

Thompson, P. B. *Food Biotechnology in Ethical Perspective.* London: Aspen Publishing, 1997.

Winston, Mark L. *Travels in the Genetically Modified Zone.* Cambridge: Harvard University Press, 2002.

Chapter Seven: Geneware Is Not Free Anymore

Andrews, Lori B. and Dorothy Nelkin. *The Body Bazaar: The Market for Human Tissue in the Biotechnology Age.* New York: Crown, 2001.

Diamond v. Chakrabarty 447 U.S. 303 (1980).

Magnus, David, Arthur Caplan, Glenn McGee, eds. *Who Owns Life?* New York: Prometheus, 2002.

Merz, J. F. "Patents Limit Medical Potential of Sequencing," *Nature*, 419(6910):878, 2002 Oct 31.

Merz, J. F. and M. K Cho. "Disease Genes Are Not Patentable: A Rebuttal of McGee," *Cambridge Quarterly of Healthcare Ethics*, 7(4):425–8, 1998 Fall.

United States Office of Technology Assessment Staff. *New Developments in Biotechnology: Patenting Life.* Washington D.C.: Marcel Dekker, 1990.

Chapter Eight: Geneware and the New Infertility

Adams, Mark B., ed. *The Wellborn Science: Eugenics in Germany, France, Brazil, and Russia.* New York: Oxford University Press, 1990.

Bridwell, Debra. *The Ache for a Child.* New York: Chariot Victor Books, 1994.

Caplan. *The Sociobiology Debate.* New York: HarperCollins, 1978.

Carter, Jean W. and Michael Carter. *Sweet Grapes: How to Stop Being Infertile and Start Living Again.* New York: Perspectives Press, 1998.

Fletcher, Joseph. *The Ethics of Genetic Control: Ending Reproductive Roulette.* New York: Prometheus, 1988.

Kevles, Daniel J. *In the Name of Eugenics: Genetics and the Uses of Human Heredity.* Los Angeles: University of California Press, 1985.

Klotzko, Arlene Judith. *The Cloning Sourcebook.* New York: Oxford University Press, 2001.

Pernick, Martin. *The Black Stork.* New York: Oxford University Press, 1999.

Silver, Lee M. *Remaking Eden: How Genetic Engineering and Cloning Will Transform the American Family.* New York: Avon, 1998.

Sutcliffe, Alastair. *IVF Children: The First Generation: Assisted Reproduction and Child Development.* London: CRC Press, 2002.

Wolpe, P. R. "If I Am Only My Genes, What Am I: Genetic Essentialism and the Jewish Response," *Kennedy Institute Journal of Ethics,* 1998.

acknowledgments

This book is the product of others' youth, a book about a matrix of new technologies that required the fire of those who can operate them and who live with them. A small army of bioinformatics, computer science, and philosophy students joined me for this project at the University of Pennsylvania Center for Bioethics. The leaders were Penn Summer Fellow Sabrina Thanse from NYU and Kelly Carroll, executive managing editor of *The American Journal of Bioethics* and coordinator of our bioethics education project. They tirelessly scoured the literature of genomics and bioinformatics, translating an impossibly complex array of problems into a matrix even I could understand. Also hard at work reading, researching, organizing, and editing were Vanessa Kuhn, Andrea Zawerczuk, Abby Horn, Andrea Gurmankin, Michelle Horowitz, and Jessica Sperling. The 1998, 2000, and 2001 participants in my state law and bioethics seminar in Penn's Department of History and Sociology of Science also helped in various ways. The digital-media director of *The American Journal*

acknowledgments

of Bioethics, bioengineer John Kwon, was there every day to answer questions and help me frame my approach to software in genomics. I had a lot of help from a bunch of people who made me feel old but seem young.

A number of colleagues helped with the development of the ideas in this volume in various ways over the past few years; they include George Agich, Anita L. Allen, Lori B. Andrews, John Arras, Charles Bosk, Dudley Burton, Tod Chambers, Ronald Dworkin, Arri Eisenstadt, Ezekiel Emanuel, Juan Enriquez, James Fowler, Mark Fox, Sen. William Frist, John Gearhart, Jonathan Glover, Peter Gorner, Hank Greely, Michael Johns, Eric Juengst, Jeffrey Kahn, Muin Khoury, George Khushf, Barbara Koenig, Mark Kuczewski, John Lachs, Paul Lanken, Richard Lewontin, David Magnus, Luigi Mastroinni, Daniel McGee, Phillip Mc-Reynolds, Jonathan Moreno, Jeff Murray, Pilar Ossorio, Eric Parens, Pasquale Patrizio, Antonio Regalado, Uwe Reinhardt, John Robertson, Mark Sauer, David Shenk, Lee Silver, Jeffrey Spike, Bonnie Steinbock, Gregory Stock, Sheryl Stolberg, Carson Strong, Rosemary Tong, Peter Ubel, Craig Venter, Nicholas Wade, Robert Weir, and Laurie Zoloth. I thank George Cunningham of the California Department of Health Services Genetic Disease Branch, Joe Hackett of the U.S. Food and Drug Administration, Paul Miller of the U.S. Equal Employment Opportunity Commission and Penn's Board of Trustees, and Harold Shapiro of Princeton University for pushing me into the middle of several of the debates discussed in this book. And I cannot say enough about the help I received from the more than one hundred participants in my Wall Street Episcopal Church/Trinity Conference Center workshops for clergy over the past year.

acknowledgments

My colleague Arthur Caplan, chair of the Department of Medical Ethics at the University of Pennsylvania, was, as always, mentoring, challenging, supportive, and insightful, as was Arthur Rubenstein, dean of our medical school; I am also thrilled to acknowledge the support of Penn's president, Judith Rodin and Provost Robert Barchi; I know it is not often that a dean, provost, and president are so supportive of such controversial teaching and work. Several of my colleagues with whom key ideas in this book were developed are owed a particular debt: Charles Bosk, David Cassarett, Mildred Cho, Renée Fox, Jason Karlawish, Donald Light, David Magnus, Jon Merz, Skip Nelson, Skip Rosoff, Pamela Sankar, and Paul Root Wolpe. Of course this doesn't mean for a moment that they are to blame for my crazy ideas or oversights.

I am grateful for financial support for my work and for elements of this project, which I received from the Greenwall Foundation, the Dodge Foundation, the Scott Foundation, the Pew Charitable Trusts, deCODE Genetics, and the United States Department of Education. I am grateful to the Foreign Office of the British Government and the British Council for my Atlantic Fellowship in Public Policy, which both enriched my opportunity to research genomics and gave me exposure to its international dimensions. In that capacity I was very fortunate to have as host King's College London School of Law and its helpful students, faculty, clergy, scientists, and clinicians. At Penn, Gloria Jones saw to it that this project made it through an extensive rewrite and an even more complex adaptation to the varied audiences of Generations X, Y, and even Z.

Portions of this book were presented as the 2001 David Smith Lecture of the Sagan Symposium at Ohio Wesleyan, as the 2001 acknowledgments

and 2002 Yale University Law and Technology Lectures, as the
2001 Addison Roache Memorial Lecture at Indiana University, as
the 2001 Carol Locke Simpson Lectures at Florida State, and as
the 2000 Henry Howe Chan Memorial Lectures at the Medical
College of Wisconsin. The discussion of mutation was presented
to the International Society for Social, Historical, and Philosophi-
cal Study of Biology in Oaxaca, Mexico. Comments, questions,
and various bursts of outrage from those in all these venues were
an important part of the development of the book you have before
you, as was response to actual chapters of the book presented at
the 2003 Munich International Conference of Stem Cells, and at
the 2003 Abu Dhabi Conference on the future of biotechnology.

This book simply would not have come into existence without the help of Rafe Sagalyn. The writing was enhanced by Jennifer Graham and my extraordinary editors at Morrow, Jennifer
Hershey and Jennifer Brehl. Each of these was skillful, visionary,
and patient in ways I could not have imagined. Cathy Trost also
provided a careful hand and a science journalist's eye for detail to
the editing and rethinking of much of the manuscript.

Friends and family were really important during the difficult
months of 2001; I think of my mom, Mike and Patricia Alberto,
Carlos Arruda, Jr., Meg Caplan, Julie Magnus, and of Jeffrey
Naser, Cynthia Shar, and Ellen Berman. My amazing boys,
Aidan, Austin, and Ethan, are so full of love and growth that I
cannot restrain my optimism about the unfolding of organic life
and the importance of new ways of thinking about human
nature. Most of all I am grateful to my wife, Monica, who had no
idea what she was getting into when we agreed that I would write
this book. Her energy, labor, love, and forgiveness made this
book—and so much more—possible.

index

Page numbers in *italic* refer to tables.

index

index

index

index

DATE DUE

Demco, Inc. 38-293